AGRO-INFORMATICS

books for everyone
PAPERBACKS
NIPA® GENX ELECTRONIC RESOURCES & SOLUTIONS P. LTD.
New Delhi-110 034

www.nipaers.com
Browse, Search, Read & Buy...
Print Books
eChapters
eBooks
Publishers
Forthcoming
Subject Catalogues
New Titles
Online Resources on:
✓ Current Affairs
✓ Reasoning, Logic & Aptitude
✓ Competitive Examinations
✓ English Language Lab
✓ Writing & Pronunciation Tools
✓ Personality Development
✓ Online Programmes for Professionals
✓ Interview Preparation
Pay Using
PhonePe
Google Pay
Paytm
PayPal
UPI
CC Avenue

AGRO-INFORMATICS

G. Vanitha
Assistant Professor, Computer Science

M. Kalpana
Assistant Professor, Computer Science

Department of Physical Sciences and Information Technology
Tamil Nadu Agricultural University, Coimbatore - 641 003

nipa® books for everyone PAPERBACKS

NIPA® GENX ELECTRONIC RESOURCES & SOLUTIONS P. LTD.
New Delhi-110 034

nipa books for everyone
PAPERBACKS

**NIPA® GENX ELECTRONIC
RESOURCES & SOLUTIONS P. LTD.**

101,103, Vikas Surya Plaza, CU Block
L.S.C. Market, Pitam Pura, New Delhi-110 034
Ph : +91 11 27341616, 27341717, 27341718
E-mail: newindiapublishingagency@gmail.com
www: www.nipabooks.com

For customer assistance, please contact
Phone: + 91-11-27 34 17 17
Fax: + 91-11- 27 34 16 16
E-Mail: feedbacks@nipabooks.com

ISBN: 978-81-96089-36-8

Composed and Designed by NIPA®.

Tamil Nadu Agricultural University

Dr. P. Murugesa Boopathi
Vice-Chancellor

COIMBATORE-641 003
TAMIL NADU, INDIA

Foreword

Tamil Nadu Agricultural University is the first to introduce Information' Technology in the field of Agriculture. The application of Information Technology in day to day activities becomes part and parcel of our routine. The application of Information Technology in the biological streams like Agriculture, Medicine and Veterinary Science has been given impetus and many of the Universities in India have made Information Technology, a compulsory subject even at under graduate level, mainly to make the life sciences dynamic and application oriented.

In this context, teaching the students with this resource material is very vital. The available resource materials support the students for self learning. The content is very simple, interesting, self explanatory and easy to follow and comprehend the subject thoroughly.

The book covers the diverse areas ranging from artificial intelligence, artificial neural networks, decision support system, expert system, geographic information system, information system related to agriculture, genetic algorithm, programming language with backend tool to develop softwares.

Under this context, the book **"Agro-Informatics"** written by Mrs. G. Vanitha, Assistant Professor of Computer Science, and Mrs. M. Kalpana, Assistant Professor of Computer Science working in the Department of Physical Sciences & Information Technology are highly appropriate, timely and relevant.

On the whole the book is well written and provides many new insights. The authors deserve appreciation for bringing out a valuable book encompassing all the essential issues and subjects of Agroinformatics. The authors certainly have to be complemented for their effort.

(P. Murugesa Boopathi)

Preface

Agroinformatics allows us to cope with the information deluge in the agriculture sector in increasingly efficient and effective ways. Agroinformatics is basically a multidisciplinary subject which covers a wide spectrum of Agricultural issues and it generally draws the concepts and methods of Computer Science. It brings the application in agriculture with innovative ideas, techniques and scientific knowledge to expand the horizons of the Computer Science.

The book contains the details about the information technology applied to management and analysis of agricultural data. The book covers the diverse areas ranging from artificial intelligence, artificial neural networks, decision support system, expert system, geographic information system, information system related to agriculture, genetic algorithm, programming language with backend tool to develop softwares. The book contains fifteen chapters that highlight and discuss the various dimensions of Agroinformatics.

It is hoped that the book will provide the basic and fundamental knowledge of understanding the concepts of Bioinformatics. This book is prepared, taking into consideration the changing needs of the undergraduate curricula of various universities involved in offering courses on Agroinformatics. The book will also be a guide for the researchers. The book would also provide the required computer knowledge to the students of State Agricultural Universities and other ICAR Institutions to sharpen their understanding of Agroinformatics.

G.Vanitha
M.Kalpana

Acknowledgement

We take this opportunity to thank our Vice Chancellor, Dr. P.Murugesa Boopathi, Tamil Nadu Agricultural University, Coimbatore for having provided the necessary working atmosphere, conductive environment and permission to bring out this valuable edition.

We would like to express our heartiest thanks to Dr. S. Santhana Bosu, Dean (Engg) for his continuous encouragement and well wishes.

We wish to express our regards to Dr. R. Palanisamy, Professor & Head, Department of Physical Sciences & Information Technology for his cooperation to bring this publication in time.

We would like to thank all those individuals who have directly or indirectly contributed their time and energy to bring this edition successfully.

We thank New India Publishing Agency (New Delhi) for their efforts in bringing out this book in record time.

We invite suggestions and critical comments from the reader for further improvement and refinement of this book.

G.Vanitha
M.Kalpana

Contents

CHAPTER 1

Introduction to Agroinformatics

What is Agroinformatics?

The term Agroinformatics can be considered as information technology applied to the management and analysis of agricultural data. In context of the plant genome initiatives, the term is applied to the computational manipulation and analysis of biological sequence data.

Agroinformatics - Definitions

The term was coined in the mid-1980s. Agroinformatics is an integration of computer, mathematical and statistical methods to manage and analyze the agricultural information.

Agroinformatics is the science of developing computer databases and algorithms for purpose of speeding up and enhancing agricultural research.

It is the branch of agriculture that uses computational tools for problem solving needs of agroinformatics.

Agroinformatics is the integration of agricultural & information technology. In general, it is a combination of

- Computer science,
- Information technology,
- Agricultural and
- Genetic engineering to determine and analyze genetic information of plants.

Objectives of Agroinformatics

- Accomplishing Information Technology for agriculture
- To enable the discovery of the agricultural insights
- To find the solution for various challenges in the agricultural fields.

Information Systems Terminology

Data

Data is raw. It simply exists and has no significance beyond its existence. It can exist in any form, usable or not. It does not have any meaning of itself.

Information

Information is data that has been given meaning by the way of relational connection. The meaning given to data by the way in which it is interpreted.

Knowledge

Knowledge is the appropriate collection of information such that its intent is to be useful. Knowledge is information with guidance for action based upon insight and experience.

Understanding

Understanding is an analytical and probabilistic process. It is the process by which the new knowledge is synthesized from the previously held knowledge. The difference between understanding and knowledge is the difference between "learning" and "memorizing".

Wisdom

Wisdom is an extrapolative and non-deterministic, non-probabilistic process. The ability or result of an ability to think, utilize knowledge, experience, understanding, common sense and insight.

Agroinformatics Tools

Various Agroinformatics tools are as follows:

- Agriculture Information System (AIS)
- Farm Management System
- Management Information Systems (MIS)
- Geographical Information Systems (GIS) and Remote Sensing
- Artificial Intelligence and Intelligent technologies in Agriculture.
- Artificial Neural Networks (ANN)
- Expert systems
- Decision Support Systems(DSS)
- Genetic Algorithms
- Programming Languages - Visual Basic
- Database Management Systems (DBMS) -SQL
- Agricultural Databases.

These tools can provide the solutions to many problems in agriculture.

Exercise to Practice

1. Define Agroinformatics.
2. List the various tools used for Agroinformatics
3. What is meant by data?
4. Define information.
5. Explain roles of an information system and its activities.
6. Write short notes on:
   Knowledge
   Understanding
   Wisdom

CHAPTER 2

Agriculture Information System

Introduction

Economics growth in larger parts of the world is dependent on the modernization of agriculture industry. The sustainable development of agriculture has been aided by recent development in Information and communication technologies. There is a necessity for information regarding the agriculture system on a regular basis.

Agriculture Information Systems – Major Trends

1. Emergence of Integratcd Agriculture Information system

Integrated Agriculture Information System is one among the emerging trends in the modern world. These systems are now employing better and new procedure such as artificial intelligence techniques, Decision Support system etc.

2. Increase in Usage of Satellite Data Transmission system

Usage pattern of the satellite data transmission system is increasing in developed countries. Data Transmission Corporation of USA manages two satellite data transmission systems in the field of agriculture namely AgDaily and FarmDayta. Utilizing satellites and leased data terminals these systems transmit data and information on various agriculture related topics.

Agriculture Information System - Importance

- Economic growth
- Agriculture industry
- Information and communication technology
- Requirement of timely information

Internet Based Agriculture Information System

Internet based agriculture information system are gaining popularity since internet is being adopted at a rapid speed.

The World Wide Web servers can provide dynamic information like weather data, current information, forecast information etc. Internet based agriculture information systems are available, the farmers can rely on an additional source of information other than local agriculture advisor, a company salesman, information newspapers and telephone services.

Support Provided by Agriculture Information System

The support provided by agriculture information system for farmers and researchers are many time helpful for very specific decisions. For example, it provides solution to protect a rice crop on a specific farm against diseases.

AGRIS/CARIS

- AGRIS is the international information system for the agricultural sciences and technology.
- It was created by the Food and Agriculture Organization of the United Nations (FAO) in 1974.

Objectives of AGRIS

- To build a common information system for science and technology in agriculture.
- AGRIS is a cooperative system in which participating countries input references and, in return, draw on the information provided by the other participants.

Principles

- International and multilingual
- A centralized collection of bibliographic details of outputs and activities of national agricultural research programmes.
- Supported by FAO
- Common ownership and free access

Major Observations

- Difficult access to the original document
- Incomplete coverage
- Independent systems

AGRIS Databases

- New AGRIS Search Engine - databases
- CARIS - databases
- FAO On-line Catalogue - databases
- FAO Resource Finder – databases
- Bibliographic databases
- Project Databases

AGROVOC

AGROVOC is a multilingual structured and controlled vocabulary designed to cover the terminology of all subject fields in agriculture, forestry, fisheries, food and related domains (e.g. environment).

- The AGROVOC Thesaurus was developed by FAO and Commission of the European Communities in the early 1980s. AGROVOC is being updated continuously by FAO.
- AGROVOC is mainly used for indexing and retrieving data in agricultural information systems both inside and outside FAO.
- AGROVOC is available in 16 languages: the five FAO official languages (which are English, French, Spanish, Chinese and Arabic), Czech, Portuguese, Japanese, Thai, Slovak, German, Hungarian, Polish, Farsi, Italian and Hindi.

Agriculture Research Information System (ARIS)

The project "Agricultural Research Information System (ARIS)" is being implemented to bring information for National Agricultural Research System (NARS).

ARIS Modules

ARIS has four information modules namely

- Agricultural Research Personnel Information System (ARPIS)
- Agricultural Research Financial Information System (ARFIS)
- Agricultural Research Library Information System (ARLIS) and
- Agricultural Research Management Information System (ARMIS).

Agriculture Research Information Center

The ICAR is also running an Agricultural Research Information Centre (ARIC) as the Central source of information on all research projects and schemes financed by the ICAR.

ARIC → AGRIS/CARIS

The ARIC is the national input centre for the AGRIS and CARIS agricultural databases of the FAO, which is the largest information system of its kind in the world.

Exercise to Practice

1. What are the major trends in Agricultural Information system?
2. Give in detail about the importance of Agricultural Information System.
3. Write about AGRIS/CARIS.
4. Explain the objective and principle of AGRIS/CARIS.
5. Write in detail about AGROVOC
6. Expand ARIS.
7. What are some of the areas in which information system based innovation have affected major functional areas of agriculture?

CHAPTER 3

Farm Management System

Introduction

The term "Farm Management System" describes the various on-farm programs to assist farmers for better management of their enterprise, natural resources and farm workers.

A farm management system offers the user numerous benefits including:

- Accurate record keeping of all activities on land level or management zone
- The farmer is able to receive detailed and accurate information regarding their farming activities
- The solution allows for better decision-making and processing accurate information.

The Concept of a Farm Management System

The Farm Management System helps the farmers to identify, manage risks and opportunities arising from their farming activity. Management attention is focused on implementing recommended practices to address the identified risks and then reviewing progress made against plans and desired outcomes.

Farm Management System is a program developed or developing to suit and match industry priorities, needs and circumstances. More information on member organizations' programs is available on their websites.

The usage of Farm Software Systems is in the area of

- Financial Management
- Crop Management Systems
- Live Stock Management Systems

Special Farm management software is available in the area of

- Ration Balancing
- Irrigation Scheduling
- Pesticide selection
- Livestock culling
- Fertilizer Management etc.

Farm Files Crops is an agricultural software package that tracks harvest data, crops, chemicals, fertilizer, planting, tillage, contracts, elevator inventory, expenses, net income, custom trucking and field mapping for the farm. This farm software allows tracking important information on the farm such as the previous crop, condition of the field when planted, planting depth, planting population, rates, etc. This agricultural software package allows to generate custom-made reports for a specific person, year, field, commodity or elevator to send to tenants or landlords.

Farm Files Livestock is a Cow/Calf software package that tracks breeds, treatments, moving cattle, recording breeds, measuring weight gains, purchase and sales transactions, deaths and blood lines for your cattle.

Exercise to Practice

1. List the Software for farm management system.
2. Uses of farm files crops
3. Explain in detail about Farm Files Livestock
4. Uses of Farm Management System.

CHAPTER 4

Management Information System

Introduction

Management Information Systems (MIS) is the term given to the discipline focused on the integration of computer systems with the aims and objectives on an organization.

The development and management of information technology tools assists executives and the general workforce in performing any tasks related to the processing of information. MIS and business systems are especially useful in the business data and the production of reports to be used as tools for decision making.

It is represented diagrammatically as

Data → Information → Communication → Decisions

MIS provides managers with information and support for effective decision making and provides feedback on daily operations. Output or reports are usually generated through accumulation of transaction processing data. Each MIS is an integrated collection of subsystems which are typically organized along functional lines within an organization

Management Information Systems consist of hardware and software that are used to process information automatically. MIS systems are used within organizations to allow many individuals to access and modify information. In many organizations, the management information system mainly operates behind the scene, and the user community is rarely involved or even aware of the processes that are handled by the MIS.

A computer system used to process orders for a business could be considered as management information system because it is assisting users of the system in automating processes related to orders. Other examples of modern management information systems are websites that process transactions for an organization or even those that serve support requests to users. A simple example of a management information system might be the support website for a product could be considered an MIS because it returns information to the end user after some initial input is provided.

Online bill pay at a bank also qualifies as a management information system - when a bill is scheduled to be paid, the user has provided information for the system to act against. The management information system then processes the payment when the due date approaches. The automated action taken by the online system is to pay the bill as requested. Since the bills within an online bill pay system can be scheduled to be automatically paid month after month, the user is not required to provide further information. Many times the bill pay system will also produce an email for the user to let them know that the action has occurred and what the outcome of the action was.

Management Information Systems typically have their own staff whose function it is to maintain existing systems and implement new technologies within a company. These positions are often highly specialized, allowing a team of people to focus on different areas within the computer system.

Water Ware Management Information System is an integrated, model-based information and decision support system for water resources management. It is implemented in open, object-oriented client-server architecture, fully web-enabled and Internet based, supporting the integration of databases, GIS, simulation and optimization models, and analytical tools into a common, easy-to-use framework. This includes a multimedia user interface with Internet access, a hybrid GIS with hierarchical map layers, object databases, time series analysis, reporting functions, an embedded expert system for estimation, classification and impact assessment tasks and a hypermedia help and explain system.

Many people use management information systems every day without thinking about the actual system they are using. Sure the individual will see a website and enter information, expecting an action to happen, but these websites, just like the accounting systems used by large corporations, are in fact management information systems.

Characteristics of Management Information System

- Provides reports with fixed and standard formats
    - Hard copy and soft copy reports
- Uses internal data stored in the computer system
- End users can develop custom reports
- Requires formal requests from users

Management Information Systems Advantage

- Provides support to managers as they work to achieve corporate goals.
- Enables managers to compare results to established company goals and identify problem areas and opportunities for improvement.

Outputs of a Management Information System

Scheduled reports

These are produced periodically or on a schedule (daily, weekly, and monthly).

Key-indicator report

Summarizes the previous day's critical activities and typically available at the beginning of each day.

Demand report

This gives certain information at a manager's request.

Exception report

This is automatically produced when a situation is unusual or requires management action.

Exercise to Practice

1. Define Management Information System
2. What are the types of reports in management information system?
3. Explain the characteristics of management information system.
4. Uses of Water Ware Management Information System.

CHAPTER 5

Geographic Information System

Introduction

A geographic information system (GIS) is a system for creating, storing, analyzing and managing spatial data and associated attributes. In other words, it is a computer system capable of integrating, storing, editing, analyzing, sharing, and displaying geographically-referenced information. In a more generic sense, GIS is a tool that allows users to create interactive queries (user created searches), analyze the spatial information, and edit data.

Defining GIS

A GIS is an information system designed to work with data referenced by spatial/ geographical coordinates. GIS is both a database system with specific capabilities for spatially referenced data as well as a set of operations for working with the data. It may also be considered as a higher order map.

GIS technology integrates common database operations such as query and statistical analysis with the unique visualization and geographic analysis benefits offered by maps. These abilities distinguish GIS from other information systems and make it valuable to a wide range of public and private enterprises for explaining events, predicting outcomes, and planning strategies.

A Geographic information system is a computer based system which is used to digitally reproduce and analyze the feature present on earth surface and the events. A typical GIS can be understood by the help of various definitions given below:-

- A Geographic information system (GIS) is a computer-based tool for mapping and analyzing things that exist and events that happen on Earth..
- GIS is a "Set of tools for collecting, storing, retrieving at will, transforming and displaying spatial data from the real world for a particular set of purposes".
- GIS is, "a computer based system that provides four sets of capabilities to handle geo-referenced data :
  1. data input
  2. data management (data storage and retrieval)
  3. manipulation and analysis
  4. data output. "

Fundamentals of GIS

Mapping Concepts, Features and Properties

A map represents geographic features or other spatial phenomena by graphically conveying information about locations and attributes. Location information describes the position of particular geographic features on the Earth's surface, as well as the spatial relationship between features, such as the shortest path from a fire station to a library, the proximity of competing businesses, and so on.

Attribute information describes characteristics of the geographic features represented, such as the feature type, its name or number and quantitative information such as its area or length. Thus the basic objective of mapping is to provide

- descriptions of geographic phenomenon
- spatial and non spatial information
- map features like Point, Line & Polygon.

Geographical Data Sets

Geographical Data Types

Although the two terms, data and information, are often used indiscriminately, they both have a specific meaning. Data can be described as different observations, which are collected and stored. Information is that data, which is useful in answering queries or solving a problem. Digitizing a large number of maps provides a large amount of data after hours of works, but the data can only render useful information if it is used in analysis.

Spatial & Non-spatial Data

Geographic data are organized in a geographic database. This database can be considered as a collection of spatially referenced data that acts as a model of reality. There are two important components of this geographic database: its geographic position and its attributes or properties data.

Attribute Data

The attributes refer to the properties of spatial entities. They are often referred to as non-spatial data since they do not in themselves represent location information.

District Name	Area	Population
Coimbatore	7469 Sq. km	42,71,856
Salem	5205 Sq. km	30,16,346
Erode	76.21 Sq. km	1,51,184

Spatial Data

Geographic position refers to the fact that each feature has a location that must be specified in a unique way. To specify the position in an absolute way a coordinate system is used. For small areas, the simplest coordinate system is the regular square grid. For larger areas, certain approved cartographic projections are commonly used. Internationally there are many different coordinate systems in use. Geographic object can be shown by four types of representation viz., points, lines, areas and continuous surfaces.

Point Data

Points are the simplest type of spatial data. They are-zero dimensional objects with only a position in space but no length.

Line Data

Lines (also termed segments or arcs) are one-dimensional spatial objects. Besides having a position in space, they also have a length.

Area Data

Areas (also termed polygons) are two-dimensional spatial objects with not only a position in space and a length but also a width (in other words they have an area).

Remote Sensing in Agriculture

Remote sensing is a tool which has been around for a long time but has had limited application in agriculture. Commercial remote sensing has greatly improved over the last few decades as the quality of digital image has been improved. An increasing number of growers are looking towards image as a tool to enhance their farming operation. Remote sensing offers potential for improving farm management practices coupled with geographic information systems (GIS).

Definition of Remote Sensing

Remote sensing is the ability to measure the properties of an object or area without making physical contact with the object. Sensors are used to collect data to improve our understanding of objects. The principles employed may range from acoustic to electromagnetic distributions. In agriculture, most remote sensing utilizes the electromagnetic spectrum.

Remote Sensing in Farm

When farmers observe their fields to assess their condition without physically touching them, it is a form of remote sensing. Observing the colors of leaves or the overall appearances of plants can determine the plant's condition. Remotely sensed images taken from satellites and aircraft provide a means to assess field conditions without physically touching them from a point of view high above the field.

Most remote sensors see the same visible wavelengths of light that are seen by the human eye, although in most cases remote sensors can also detect energy from wavelengths that are undetectable to the human eye. The remote view of the sensor and the ability to store, analyze, and display the sensed data on field maps are what make remote sensing a potentially important tool for agricultural producers.

Remotely sensed images can be used to identify nutrient deficiencies, diseases, water deficiency or surplus, insect damage, hail damage, wind damage, herbicide damage, and plant populations. Information from remote sensing can be used as base maps in variable rate applications of fertilizers and pesticides. Information from remotely sensed images allows farmers to treat only affected areas of a field. Problems within a field may be identified remotely before they can be visually identified.

Remote Sensing and its Importance in Agricultural Survey

In such study single tabular form of data or map data is not sufficient enough which can be combined with information's obtained from existing maps and tabular data.

- Remote Sensing techniques using various platform has provide its utility in agricultural survey.
- Satellite data provides the actual synoptic view of large are at a time, which is not possible from conventional survey methods.
- The process of data acquisition and analysis is very fast through Geographic information system (GIS) as compared to conventional methods

Remote Sensing techniques have a unique capability of recording data in visible as well as invisible (i.e. ultraviolet, reflected infrared, thermal infrared and microwave etc.) part of electromagnetic spectrum. Therefore certain phenomenon, which cannot be seen by human eye, can be observed through remote sensing techniques i.e. the trees, which are affected by disease, or insect attack can be detected by remote sensing techniques much before human eyes see them.

Principles of Remote Sensing

The basic principles of remote sensing with satellites and aircraft are similar to visual observations. Energy in the form of light waves travels from the sun to Earth. Light waves travel similarly to waves traveling across a lake. The distance from the peak of one wave to the peak of the next wave is the wavelength. Energy from sunlight is called the electromagnetic spectrum.

The wavelengths used in most agricultural remote sensing applications cover only a small region of the electromagnetic spectrum. Wavelengths are measured in micrometers (µm) or nanometers (nm).

One μm is about .00003937 inch and 1 μm equals 1,000 nm. The visible region of the electromagnetic spectrum is from about 400 nm to about 700 nm. The green color associated with plant vigor has a wavelength that centers near 500 nm.

Electromagnetic Energy and Plants

When electromagnetic energy from the sun strikes plants, three things can happen. Depending upon the wavelength of the energy and characteristics of individual plants, the energy will be reflected, absorbed, or transmitted. Reflected energy bounces off leaves and is readily identified by human eyes as the green color of plants. A plant looks green because the chlorophyll in the leaves absorbs much of the energy in the visible wavelengths and the green color is reflected. Sunlight that is not reflected or absorbed is transmitted through the leaves to the ground.

Interactions between reflected, absorbed, and transmitted energy can be detected by remote sensing. The differences in leaf colors, textures, shapes or even how the leaves are attached to plants, determine how much energy will be reflected, absorbed or transmitted. The relationship between reflected, absorbed and transmitted energy is used to determine spectral signatures of individual plants. Spectral signatures are unique to plant species.

How Does Remote Sensing Work?

There are several types of remote sensing systems used in agriculture but the most common is a passive system that senses the electromagnetic energy reflected from plants. Sun is the most common source of energy for passive systems. Passive system sensors can be mounted on satellites, manned or unmanned aircraft, or directly on farm equipment.

There are several factors to consider when choosing a remote sensing system for a particular application, including spatial resolution, spectral resolution, radiometric resolution, and temporal resolution.

Global Positioning System

Introduction

GPS stands for the Global Positioning System. It refers to a system of satellites and receivers that allow people and devices to pinpoint their precise location on the earth. Devices that are equipped with

GPS equipment receive transmissions from at least a few of the satellites and are able to discern very precise positioning data.

How does it Work?

1. The basis of GPS is "triangulation" from satellites.
2. To "triangulate," a GPS receiver measures distance using the travel time of radio signals.
3. To measure travel time, GPS needs very accurate timing which it achieves with some tricks.
4. Along with distance, we need to know exactly where the satellites are in space. High orbits and careful monitoring are the secret.
5. Finally we must correct for any delays the signal experiences as it travels through the atmosphere.

Components of a GPS

The GPS is divided into three major components

- The Control Segment
- The Space Segments
- The User Segment

The Control Segment

The Control Segment consists of five monitoring stations (Colorado Springs, Diego Garcia, Hawaii, and Kwajalein Island). Three of the stations (Ascension, Diego Garcia, and Kwajalein) serve as uplink installations, capable of transmitting data to the satellites, including new ephemerides (satellite positions as a function of time), clock corrections and other broadcast message data, while Colorado Springs serves as the master control station. Satellite tracking data from the monitoring stations are transmitted to the master control station for processing.

This processing involves the computation of satellite ephemerides and satellite clock corrections. The master station controls orbital corrections, when any satellite strays too far from its assigned position, and necessary repositioning to compensate for unhealthy (not fully functioning) satellites.

The Space Segment

The Space Segment consists of the Constellation of NAVASTAR earth orbiting satellites. The current Defence Department plan calls for a full constellation of 24 Block II satellites (21 operational and 3 in-orbit spares). The satellites are arrayed in 6 orbital planes, inclined 55 degrees to the equator. They orbit at altitudes of about 12000, miles each, with orbital periods of 12 sidereal hours (i.e., determined by or from the stars), or approximately one half of the earth's periods, approximately 12 hours of 3-D position fixes. The next block of satellites is called Block IIR, and they will provide improved reliability and have a capacity of ranging between satellites, which will increase the orbital accuracy. Each satellite contains four precise atomic clocks and has a microprocessor on board for limited self-monitoring and data processing. The satellites are equipped with thrusters which can be used to maintain or modify their orbits.

The User Segment

The user segment is a total user and supplier community, both civilian and military. The User Segment consists of all earth-based GPS receivers. Receivers vary greatly in size and complexity, though the basic design is rather simple. The typical receiver is composed of an antenna and preamplifier, radio signal microprocessor, control and display device, data recording unit, and power supply. The GPS receiver decodes the timing signals from the 'visible' satellites (four or more) and, having calculated their distances, computes its own latitude, longitude, elevation, and time. This is a continuous process and generally the position is updated on a second-by-second basis, output to the receiver display device and, if the receiver display device and, if the receiver provides data capture capabilities, stored by the receiver-logging unit.

GPS Positioning Types

Absolute Positioning

The mode of positioning relies upon a single receiver station. It is also referred to as 'stand-alone' GPS, because, unlike differential positioning, ranging is carried out strictly between the satellite and the receiver station, not on a ground-based reference station that assists with the computation of error corrections. As a result, the positions derived in absolute mode are subject to the unmitigated errors inherent in satellite positioning. Overall accuracy of absolute positioning is considered to be no greater than 50 meters at best by Ackroyd and Lorimer and to be + 100 meter accuracy by the U.S. Army Corps of Engineers.

Differential Positioning

Relative or Differential GPS carries the triangulation principles one step further, with a second receiver at a known reference point. To further facilitate determination of a point's position, relative to the known earth surface point, this configuration demands collection of an error-correcting message from the reference receiver.

Differential-mode positioning relies upon an established control point. The reference station is placed on the control point, a triangulated position, the control point coordinate. This allows for a correction factor to be calculated and applied to other roving GPS units used in the same area and in the same time series. Inaccuracies in the control point's coordinate are directly additive to errors inherent in the satellite positioning process. Error corrections derived by the reference station vary rapidly, as the factors propagating position errors are not static over time. This error correction allows for a considerable amount of error of error to be negated, potentially as much as 90 percent.

GPS Setup

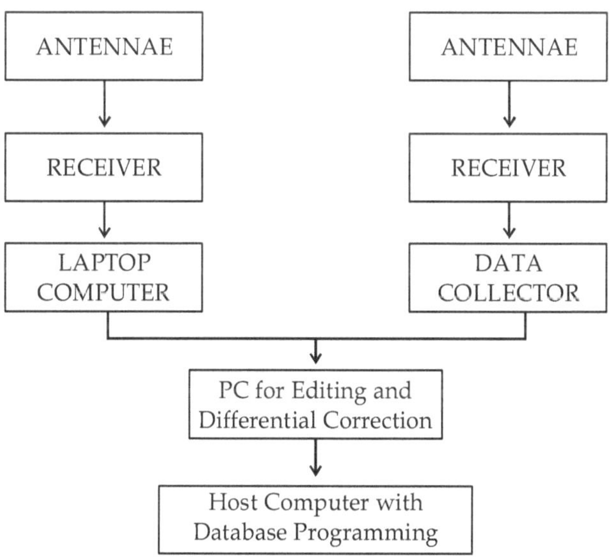

GPS Setup

Accuracy of GPS

There are four basic levels of accuracy - or types of solutions - you can obtain with your real-time GPS mining system:

Autonomous	Accuracy	15 - 100 meters
Differential GPS (DGPS)	Accuracy	0.5 - 5 meters
Real-Time Kinematic Float (RTK Float)	Accuracy	20cm - 1 meter
Real-Time Kinematic Fixed (RTK Fixed)	Accuracy	1cm - 5 cm

GPS satellites broadcast on three different frequencies, and each frequency (or career wave) has some information or codes on it. We can think of it as three different radio stations broadcasting several different programs. The table below lists the signals and the contents:

L1 Career	L2 Career	L3 Career
19 cm wavelength	24 cm wavelength	Data not available
1575.42 M Hz	1227.6 M Hz	
C/A Code	P Code	
Navigation	Navigation Message	

- P Code : Reserved for direct use only by the military
- C/A Code : Used for rougher positioning
- For Single frequency use only L1 career is used
- For Double frequency, L1/L2/L3 career is used
- The navigation message (usually referred to as the ephemeris) tells us where the satellites are located, in a special coordinate system called WGS-84. If we know where the satellites are at any given time, then we can compute your location here on earth.

Factors that affect GPS

There are a number of potential error sources that affect either the GPS signal directly or your ability to produce optimal results:

Number of satellites - minimum number required: You must track atleast four common satellites - the same four satellites - at both the reference receiver and rover for either DGPS or RTK solutions. Also to achieve centimeter -level accuracy, remember we must have a fifth satellite for on-the fly RTK initialization. This extra satellite adds a check on the internal calculation. Any additional satellites beyond five provide even more checks, which is always useful.

- Multipath - reflection of GPS signals near the antennae: Multipath is simply reflection of signals similar to the phenomenon of ghosting on our television screen. GPS signals may be reflected by surfaces near the antennae, causing error in the travel time and therefore error in the GPS positions.
- Ionosphere - change in the travel time of the signal: Before GPS signals reach antenna on the earth; they pass through a zone of charged particles called the ionosphere, which changes the speed of the signal. If reference and rover receivers are relatively close together, the effect of ionosphere tends to be minimal. And if we are working with the lower range of GPS precisions, the ionosphere is not a major consideration. However if our rover is working too far from the reference station, we may experience problems, particularly with initializing your RTK fixed solution.
- Troposphere - change in the travel time of the signal: Troposphere is essentially the weather zone of our atmosphere, and droplets of water vapour in it can affect the speed of the signals. The vertical component of our GPS answer (our elevation) is particularly sensitive to the troposphere.
- Satellite Geometry - general distribution of the satellites: Satellite Geometry - or the distribution of satellites in the sky - effects the computation of your position. This is often referred to as Position Dilution of Precision (PDOP).

GPS Applications

One of the most significant and unique features of the Global Positioning Systems is the fact that the positioning signal is available to users in any position worldwide at any time. With a fully operational GPS system, it can be generated to a large community of likely to grow as there are multiple applications, ranging from surveying, mapping, and navigation to GIS data capture. The GPS will soon be a part of the overall utility of technology. There are countless GPs applications, a few important ones are covered in the following passage.

Surveying and Mapping

The high precision of GPS carrier phase measurements, together with appropriate adjustment algorithms, provides an adequate tool for a variety of tasks for surveying and mapping. Using DGPs methods, accurate and timely mapping of almost anything can be carried out. The GPS is used to map cut blocks, road alignments, and environmental hazards such as landslides, forest fires, and oil spills. Applications, such as cadastral mapping, needing a high degree of accuracy also can be carried out using high grade GPS receivers. Continuous kinematic techniques can be used for topographic surveys and accurate linear mapping.

Navigation

Navigation using GPS can save countless hours in the field. Any feature, even if it is under water, can be located up to one hundred meters simply by scaling coordinates from a map, entering waypoints, and going directly to the site. Examples include road intersections, corner posts, plot canters, accident sites, geological formations, and so on. GPS navigation in helicopters, in vehicles, or in a ship can provide an easy means of navigation with substantial savings.

Remote Sensing and GIS

It is also possible to integrate GPS positioning into remote-sensing methods such as photogrammetry and aerial scanning, magnetometry, and video technology. Using DGPS or kinematic techniques, depending upon the accuracy required, real time or post-processing will provide positions for the sensor which can be projected to the ground, instead of having ground control projected to an image. GPS are becoming very effective tools for GIS data capture. The GIS user community benefits from the use of GPS for locational data capture in various GIS applications. The GPS can easily be linked to a laptop computer in the field, and, with appropriate software, users can also have all their data on a common base with every little distortion. Thus GPS can help in several aspects of construction of accurate and timely GIS databases.

Geodesy

Geodetic mapping and other control surveys can be carried out effectively using high-grade GPs equipment. Especially when helicopters were used or when the line of sight is not possible, GPS can set new standards of accuracy and productivity.

Military

The GPS was primarily developed for real time military positioning. Military applications include airborne, marine, and land navigation. Future of GPS Technology barring significant new complications due to S/A (Selective Availability) from DOD, the GPS industry is likely to continue to develop in the civilian community. There are currently more than 50 manufacturers of GPs receivers, with the trend continuing to be towards smaller, less expensive, and more easily operated devices.

While highly accurate, portable (hand-held) receivers are already available, current speculation envisions inexpensive and equally accurate 'wristwatch locators' and navigational guidance systems for automobiles. However, there is one future trend that will be very relevant to the GIS user community, namely, community base stations and regional receive networks, as GPS management and technological innovations that will make GPS surveying easier and more accurate.

Use of GPS in Agriculture

Tractor Guidance

Farmers cannot put their tractors on auto-pilot. If they plow their fields with a recording GPS system the tractor can then be programmed to follow the same route - for cultivating, fertilizing, pest control and harvesting. The programming of tractor routes has the potential to save a lot of money.

Crop Duster Targeting

Insects don't attack a field with a uniform distribution. Instead outbreaks of insect activity are concentrated in certain areas. Workers strolling the crops can use a GPS to record the locations of insect problems. The data can then be used by cropduster pilots to selectively target the problem areas instead of treating an entire field. This method results in a savings of time, fuel, insecticide and crop exposure to chemicals.

Tracking Livestock

The location of valuable animals on a large farm can be monitored by GPS transmitters attached to the animals collar. When the animals are sent to market GPS transmitters can also be used to track their location.

Yield Monitoring

Estimates of yield variations across a property can be made using GPS. To do this the property is divided into zones and the yield of each zone is estimated and plotted on a map. The map can then be used to better understand the property and for decision-making in regard to the next planting.

Soil Sampling

Collecting soil samples across a large property can be organized using GPS and mapping software. The sample locations can be way pointed in the field and those waypoints marked on the mapping software. Then, when the laboratory results are returned the data can be plotted on the maps and decisions for soil treatment can be made for various parts of the property. The location information can save money and time by allowing variable rate applications and treating only those areas with a documented need.

The development and implementation of precision agriculture or site-specific farming has been made possible by combining the Global Positioning System (GPS) and geographic information systems (GIS). These technologies enable the coupling of real-time data collection with accurate position information, leading to the efficient manipulation and analysis of large amounts of geospatial data. GPS-based applications in precision farming are being used for farm planning, field mapping, soil sampling, tractor guidance, crop scouting, variable rate applications, and yield mapping. GPS allows farmers to work during low visibility field conditions such as rain, dust, fog, and darkness.

In the past, it was difficult for farmers to correlate production techniques and crop yields with land variability. This limited their ability to develop the most effective soil/plant treatment strategies that could have enhanced their production.

Today, more precise application of pesticides, herbicides, and fertilizers, and better control of the dispersion of those chemicals are possible through precision agriculture, thus reducing expenses, producing a higher yield, and creating a more environmentally friendly farm.

Precision agriculture is now changing the way farmers and agribusinesses view the land from which they reap their profits.

Precision agriculture is about collecting timely geospatial information on soil-plant-animal requirements and prescribing and applying site-specific treatments to increase agricultural production and protect the environment. Where farmers may have once treated their fields uniformly, they are now seeing benefits from micromanaging their fields. Precision agriculture is gaining in popularity largely due to the introduction of high technology tools into the agricultural community that are more accurate, cost effective, and user friendly. Many of the new innovations rely on the integration of on-board computers, data collection sensors, and GPS time and position reference systems.

Many believe that the benefits of precision agriculture can only be realized on large farms with huge capital investments and experience with information technologies. Such is not the case. There are inexpensive and easy-to-use methods and techniques that can be developed for use by all farmers. Through the use of GPS, GIS, and remote sensing, information needed for improving land and water use can be collected. Farmers can achieve additional benefits by combining better utilization of fertilizers and other soil amendments, determining the economic threshold for treating pest and weed infestations, and protecting the natural resources for future use.

GPS equipment manufacturers have developed several tools to help farmers and agribusinesses become more productive and efficient in their precision farming activities. Today, many farmers use GPS-derived products to enhance operations in their farming businesses. Location information is collected by GPS receivers for mapping field boundaries, roads, irrigation systems, and problem areas in crops such as weeds or disease. The accuracy of GPS allows farmers to create farm maps with precise acreage for field areas, road locations and distances between points of interest. GPS allows farmers to accurately navigate to specific locations in the field, year after year, to collect soil samples or monitor crop conditions.

Crop advisors use rugged data collection devices with GPS for accurate positioning to map pest, insect, and weed infestations in the field. Pest problem areas in crops can be pinpointed and mapped for future management decisions and input recommendations. The same field data can also be used by aircraft sprayers, enabling accurate swathing of fields without use of human "flaggers" to guide them. Crop dusters equipped with GPS are able to fly accurate swaths over the field, applying chemicals only where needed, minimizing chemical drift, reducing the amount of chemicals needed, thereby benefiting the environment. GPS also allows pilots to provide farmers with accurate maps.

Farmers and agriculture service providers can expect even further improvements as GPS continues to modernize. In addition to the current civilian service provided by GPS, the United States is committed to implementing a second and a third civil signal on GPS satellites. The first satellite with the second civilian signal was launched in 2005. The new signals will enhance both the quality and efficiency of agricultural operations in the future.

General Packet Radio Service (GPRS) Network

What is GPRS?

The General Packet Radio Service (GPRS) network is an "always on", private network for data. It uses the existing GSM network to transmit and receive TCP/IP based data to and from GPRS mobile devices. Private IP addresses are typically dynamically assigned within the network to mobile devices. However, Access Point Names (APN's) provide a gateway route to other networks such as the Internet, WAP services or private corporate networks. Firewalls typically reside at the APN to isolate the public and private networks. IP addresses allocated to mobile GPRS devices are therefore not addressable from outside the GPRS network (e.g. from the Internet) without specialised services or infrastructure.

The GPRS Network

Public Internet Access to GPRS Devices

GPRS devices are not addressable from the internet. The Internet and GPRS are designed for client driven applications. Due to APN firewalls, remote GPRS server devices can only be addressed by a private network solution or via more complex VPN technology.

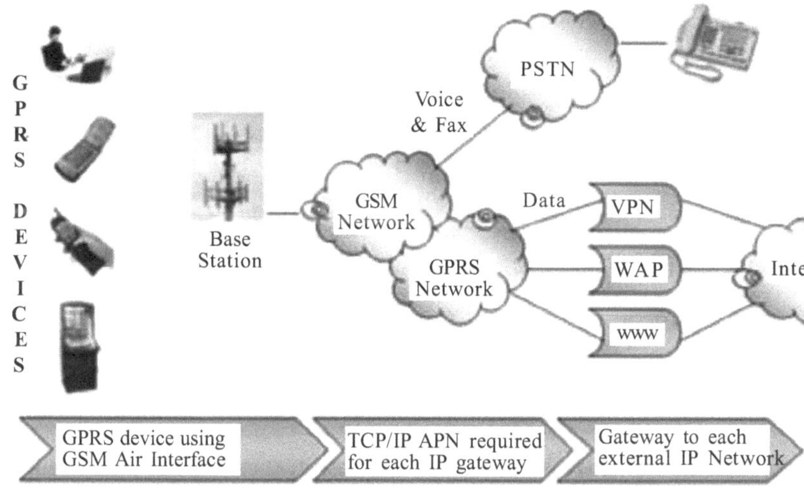

Private Network Access to GPRS Devices

Although client GPRS devices can communicate with ease over Public and Private networks, GPRS server devices require a static IP address. Network Operators offer private APN's to corporate networks over Leased Lines or VPN's, where IP address assignment is managed by the customer's corporate Network e.g. using a radius server. Alternatively, Wireless Operators in some countries offer private APN's with static IP address support thereby creating customers their own private network within the GPRS network.

GPRS Mobile Device

There are a number of GPRS devices, each of which can offer GSM services too, such as voice calls and SMS.

- A GPRS Mobile Phone
- A GPRS Radio Card for a PC
- A Hand held PC with an in-built GPRS Mobile
- A remote machine

GPRS Classes

There are 3 classes of GPRS device being developed, of which only class B is currently available: -

- *Class A* - Operates in GSM and GPRS modes at the same time, and hold simultaneous voice and data sessions.
- *Class B* - Operates in GSM and GPRS modes at the same time, and but cannot hold simultaneous calls.
- *Class C* - Can be active in either GSM or GPRS mode, but not at the same time

Benefits

- *Efficient* - GPRS mobile devices only use the GSM network when data is transferred. The GSM connection is not dedicated to each user, therefore it can be shared with many users resulting in efficient use of the network.
- *Fast* - GPRS gives speeds of upto 5 time faster than GSM. GPRS offers maximum data rates of 56Kbps (down) and 14.4kbps (up), however this is shared bandwidth therefore actual data rates are potentially lower.
- *Payment based on data usage :* Billing is not based on time, but on the amount of data actually transferred.

Exercise to Practice

1. What is the importance of having a GIS?
2. What various kinds of information are available through GIS?
3. What is the difference between GIS and a map?
4. How can GIS be used to make decisions?
5. What is the difference between GIS and surveying?
6. How can GIS help our city maintain its water system?
7. What is GPS?
8. What are some World Wide Web URLs for GIS information?
9. Describe the difference between raster and vector data models used to represent maps in a GIS.

CHAPTER 6

Artificial Intelligence

What is Intelligence?

Intelligence is the computational part of the ability to achieve goals in the world.

What is Artificial Intelligence?

Artificial Intelligence is a set of software tools that enables computers to think like humans to solve problems.

Artificial Intelligence

The branch of computer science concerned with making computers behave like humans. Artificial intelligence includes

- *Games playing*: programming computers to play games such as chess and checkers.
- *Expert systems* : programming computers to make decisions in real-life situations (for example, some expert systems help doctors diagnose diseases based on symptoms).

- *Natural language* : programming computers to understand natural human languages.
- *Neural networks* : Systems that simulate intelligence by attempting to reproduce the types of physical connections that occur in animal brains.
- *Robotics* : programming computers to see, hear and react to other sensory stimuli.

Currently, no computers exhibit full artificial intelligence (that is, are able to simulate human behavior). In the area of robotics, computers are now widely used in assembly plants, but they are capable only of very limited tasks. Robots have great difficulty identifying objects based on appearance or feel, and they still move and handle objects slowly.

Branches of AI

- Logical AI
- Search
- Pattern Recognition
- Representation
- Inference etc.

Logical AI

It deals about the facts of the situation in which it must act and its goals are represented by sentences of some mathematical logical language.

Eg: Programs help doctors to diagnose disease based on symptoms.

Search

AI programs often examine large number of possibilities.

Eg: Moves in a chess game.

Pattern Recognition

When a program makes some observations of some kind, it is often programmed to compare what it sees with a pattern.

Representation

Facts about the world have to be represented in some way. Information from the earth satellite used by the optimized pest management

Inference

- From some facts, others can be inferred.
- Flight simulation

Natural - Language

Natural - language processing offers the greatest potential rewards because it would allow people to interact with computers without needing any specialized knowledge. We could simply walk up to a computer and talk to it.

Unfortunately, programming computers to understand natural languages has proved to be more difficult than originally thought. Some rudimentary translation systems that translate from one human language to another are in existence, but they are not nearly as good as human translators. There are also voice recognition systems that can convert spoken sounds into written words, but they do not understand what they are writing they simply take dictation.

In the early 1980s, expert systems were believed to represent the future of artificial intelligence and of computers in general. To date, however, they have not lived up to expectations. Many expert systems help human experts in such fields as medicine and engineering, but they are very expensive to produce and are helpful only in special situations.

Today, the hottest area of artificial intelligence is neural networks, which are proving successful in a number of disciplines such as voice recognition and natural-language processing.

There are several programming languages that are known as AI languages because they are used almost exclusively for AI applications. The two most common are LISP and Prolog.

Natural Language Processing

Natural language processing is the most challenging problem and there is no o complete solution till today. The Fourth-generation languages are closest to the natural languages. Translation systems that translate from one human language to another are in existence.

Eg: Voice recognition systems.

Natural Language Recognition

Grammar

- Grammars define the legal structures of a language.
- "parse" a sentence using a grammar to:
  - Determine whether it is grammatical.
  - Assign some useful structure/grouping to the sentence.

Syntactic Categories

- Grammars based on each word belonging to a particular category:
  - nouns
  - verbs
  - adjectives
  - adverbs
  - articles/determiners

Larger Grouping

- Noun phrase: sequence of words denoting an object. E.g., the black cow.
- Verb phrase: sequence of words denoting an action. E.g.,
  - jumps quickly
  - runs after the small cat
  - kicks the small girl with the funny teeth

Note that verb phrases may contain noun phrases.

Simple Natural Language Grammar

We can write a simple NL grammar using phrase structure rules such as the following:

- sentence → nounPhrase, verbPhrase.
- nounPhrase → article, adjective, noun.
- verbPhrase → verb, nounPhrase.

Rules define constituent structure.

Parsing

Using these rules we can determine whether a sentence is legal, and obtain its structure.

"The large cat eats the small rat".

This consists of

- *Noun Phrase*: The large cat
- *Verb Phrase*: eats the small rat

The verb phrase in turn consists of

- *verb*: eats
- *Noun Phrase*: the small rat

Intelligent Technologies in Agriculture

The role of world agriculture will become increasingly crucial in forthcoming decades, as concerns over food, the environment, and energy increase. On-going investigations are attempting to understand, explore, and solve these issues as they relate to agriculture. The inherently complex, dynamic, and non-linear nature of agricultural systems has always required solutions based on advanced techniques and technologies to provide greater accuracy, better understanding and appropriate solutions. The recent application of the technologies of artificial intelligence (i.e. computer vision, robotics and control systems, expert systems, decision support systems, natural language processing, etc.) and other advanced forms of information technology (neural networks, fuzzy logic, genetic algorithms, and photosynthesis algorithms) promises to provide solutions to problems in agricultural systems. The promotion of these technologies as their costs decrease is catalyzing additional investigations of the application of different forms of AI technology in agriculture.

Intelligent technology is a kind of artificial simulation, which simulates a human being's activity of thinking. Expert system techniques, computing intelligence, machine learning, knowledge discovery etc are examples of intelligent technologies. Now agriculture industry view intelligent technology with optimism.

The agriculture production system is a giant complex system. It is influenced by soil, species, climate and many other factors. The application in agriculture domain has broad prospects.

Intelligent technologies can be applied to areas such as irrigation, fertilization, planting, diagnosis and prevention of diseases and insect pests.

In complex agriculture system, the knowledge and models of agriculture are usually of limitations and incomplete. It is quite effective way that machine learning (ML) and knowledge discovery in database (KDD) are used to discover laws and acquire knowledge from data and cases.

Scientists have applied the inductive learning method to acquire diagnosis rules of diseases and insect pests and utilize knowledge discovery to discover the related laws of crops growth and employed neural networks for knowledge discovery such as, prediction of rainfall, extraction of evaluation rules of soil fertility etc.

Intelligent Animal Houses

There is a growing need for computer based methods for herd management within pig production. The average herd size in pig production is increasing as well as the number of animals managed per person process computers known from factory automation, used together with electronic sensors can easily be set up at farm level to provide real time data on a personal computer.

Robotics

Robots are devices that can move and react to sensory input. The field of computer science and engineering concerned with creating robots is Robotics. Robotics is one branch of artificial intelligence.

Uses

Robots are now widely used in

- Factories to perform high-precision jobs such as welding and riveting.

Special situations that would be dangerous for humans - example, in cleaning toxic wastes or defusing bombs.

Agricultural Robots

Agricultural robots capable of collecting crop and soils samples but are now able to navigate equipment through GPS eliminating the need for a driver.

Agricultural Robotics is the logical proliferation of automation technology into biosystems such as agriculture, forestry, fisheries, etc. Robots have the advantage of being small, light weight and autonomous. Because of their size they can collect data in close proximity to the crop and soil.

Cameras are used to detect weeds, and larger scale sensors are being developed to detect crop stresses and disease.

The second generation of robots will be able to perform field operations such as mechanical weeding and micro spraying. All operations that are currently performed in the field can be done with robots, preferably of smaller size and lower cost.

The third generation of robots will be as part of a fully autonomous crop production system. Sensors will provide real time information about the status of the crop and computer software and data fusion techniques will help to digest the data into management decisions.

Autonomous Robots (Ag Robots)

These Ag Robots will be able to sample soil, detect weeds, insects, and diseases, weed, and spray pesticides or herbicides. Unlike large tractors, Ag Robots won't compact the soil as they move through the rows of crops because of their light weight. Cameras are presently used to detect weeds but larger sensors to detect crop stress and disease, insect activity sensors, and soil sensors still need to be developed. Once the sensors are developed, the next step will be to develop Ag Robots that can weed, spray precise amounts of chemicals, and plant. The goal is to eventually have an army of robots that can perform all farming activities through a system of sensors, wireless communication, and data managing software

AgBo

Army of AgAnts

AgBo is used for steering which allows for crabbing, spin turns, front/rear and all-wheel steering. The Ag Ant is a fully autonomous robot that will eventually work as part of a group of other Ag Ants to detect weeds, spray pesticides, and do soil tests. When an Ag Ant detects a weed, it will be able to communicate to the other Ag Ants to come and help destroy it.

AgTracker

AgTracker, which can turn well in muddy soil because of skid steering. It is powered by two motors and has 8 infrared sensors, ultrasound sensors, and an electronic compass. AgTracker is a much simpler robot than AgBo, it was targeted at low cost. The robot is very stable and turns well in soils even under muddy conditions.

Fruit Picking Robot

These principles of Fruit picking robot is to open new approaches to harvest crops. Fruit picking robotics technology is contributed from high-tech industry, agricultural commodity groups and farm equipment manufacturers. To start with, the fruit picking robots need to pick ripe fruit without damaging the branches or leaves of the tree. Mobility is a priority, and the robots must be able to access all areas of the tree being harvested. It goes then without saying that the robots must be 'intelligent', and have a human-like interaction with their surroundings through senses of touch, sight, and image processing. The fundamental blocks of these robots are shown in the diagram here.

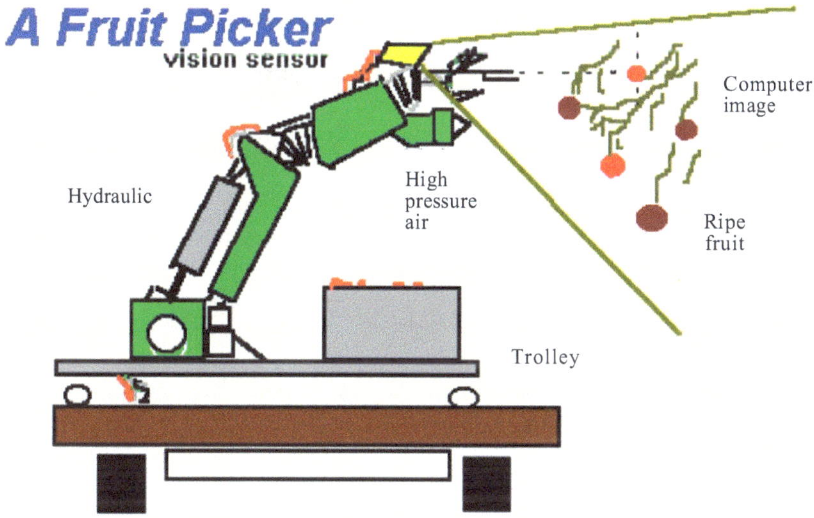

The robot can distinguish between fruit and leaves by using video image capturing. The camera is mounted on the robot arm, and the colors detected are compared with properties stored in memory. If a match is obtained, the fruit is picked.

If fruit is hidden by leaves, an air jet can be used to blow leaves out the way so a clearer view and access can be obtained. The robot arm itself is coated in rubber to minimize any damage to the tree. The pressure applied to the fruit is sufficient for removal from the tree, but not enough to crush the fruit. This is accomplished by a feedback process from the gripper mechanism, which is driven by motors, hydraulics, or a pneumatic system.

The shape of the gripper depends on the fruit being picked, as some fruits, such as plums, crush very easily, while others, like oranges are not so susceptible to bruising. The robots should have access to all areas of the orchard in order to reach all of the fruit.

Exercise to Practice

1. What is Intelligence?
2. What is artificial intelligence?
3. Explain in detail about natural language.
4. Define robotics.
5. Uses of robotics in agriculture.
6. Write about the application of intelligent technologies in agriculture.
7. Explain in detail about fruit picking robots.

CHAPTER 7

Artificial Neural Networks

What is a Neural Network?

A type of artificial intelligence that attempts to imitate the way a human brain works. A neural network works by creating connections between processing elements, the computer equivalent of neurons.

How does an Artificial Neural Network (ANN) Work?

- Neural networks process information in a similar way the human brain does.
- The network is composed of a large number of highly interconnected processing elements working in parallel to solve a specific problem.
- Neural networks learn by example.
- The examples must be selected carefully.
- The disadvantage is that because the network finds out how to solve the problem by itself, its operation can be unpredictable.

Human Brain

The human brain contains about 10 billion nerve cells (neurons). Each neuron is connected to the others through 10000 synapses. A neuron is composed of a nucleus - a cell body known as soma. Attached to soma are long irregularly shaped filaments called dendrites. Another type of attachment to soma is axon. Axon appears on the output cells. A neuron has

- Input (dendrites)
- Output (the axon)

The information circulates from the dendrites to the axon via the cell body. Axon connects to dendrites via synapses.

Human brain and neural networks

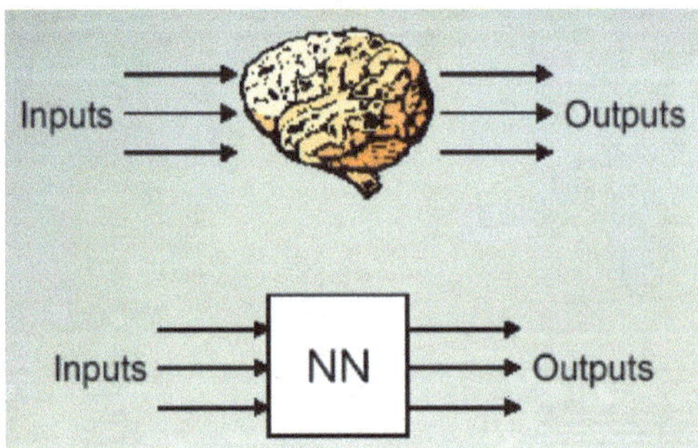

Characteristics of Neural Networks

- The NNs exhibit mapping capabilities, that is, they can map input patterns to their associated output patterns

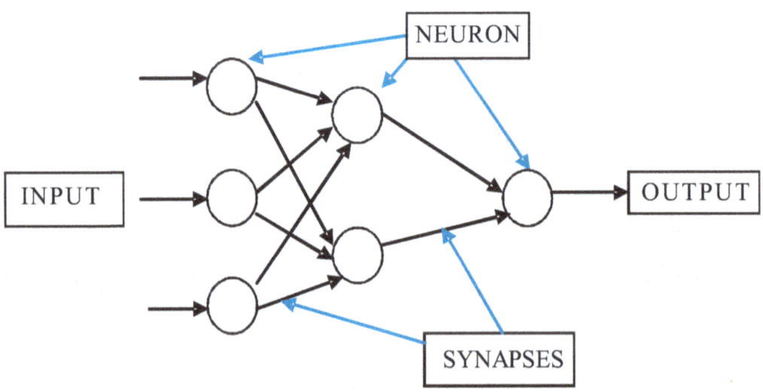

- The NNs posses the capability to generalize. Thus they can predict new outcomes from past trends.
- The NNs are robust systems and are fault tolerant. They can, therefore, recall full patterns from incomplete, partial or noisy patterns.
- The NNs can process information in parallel, at high speed and in distributed manner.

Architectures of Artificial Neural Networks

Network architecture is specified by the number of inputs to the network, the number of outputs, the total number of elementary nodes that are usually equal processing elements for the entire networks and their organization and interconnection.

Neural networks are classified into 2 categories are

1. Feedforward
   a) Single layer feedforward Network
   b) Multiple layer feedforward network
2. Recurrent network

Feed-forward networks

Information flow is unidirectional. Without any loop or feedbacks. There is no link between nodes in same layer. Outputs of nodes in a specified layer are always connected as input to nodes in succeeding layers. Data is presented to Input layer passed on to Hidden Layer passed on to Output layer and information is distributed.

Single layer Feed Forward Network

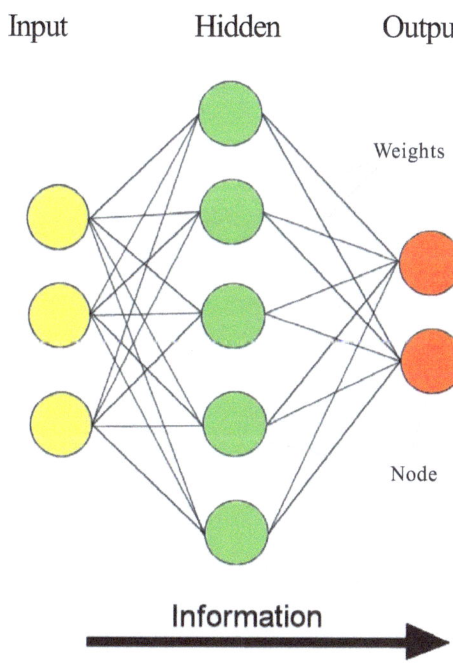

Multilayer Feed-Forward ANNs

- Signals flow from the inputs to the outputs of the network through several layers of neurones.

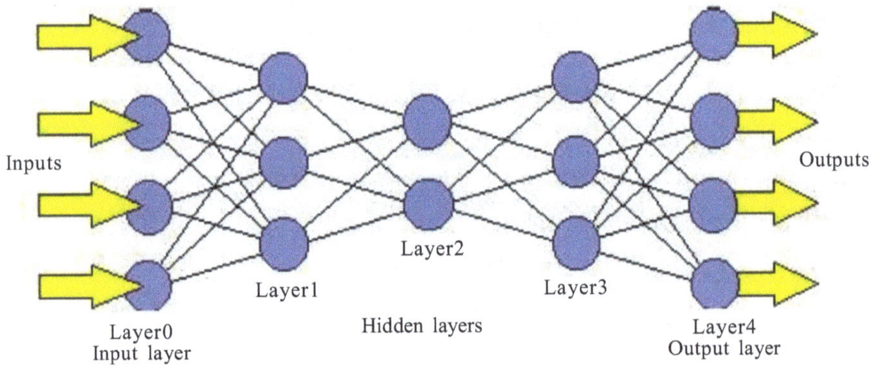

Recurrent Networks

If the feedback link forms a circular path in a network then the network is recurrent. There may be atleast one feedback link.

Hopfield Networks

If the initial network had a single neurone layer then the result is a fully connected (Hopfield) neural network.

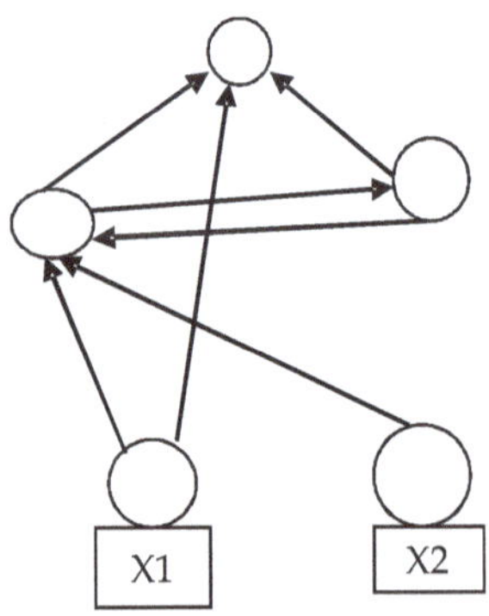

Neuron with self feedback links

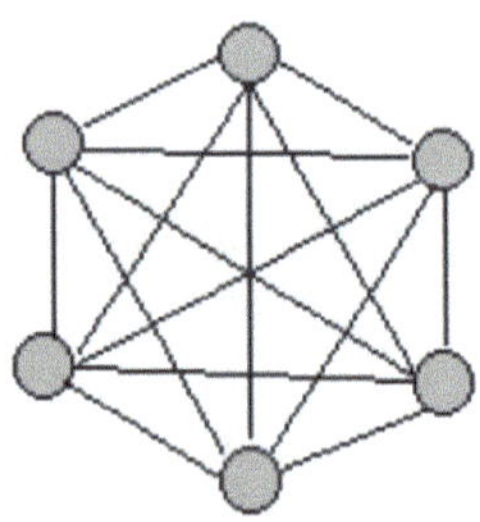

Hopfield Network

Learning Process

Learning is a process by which the free parameters of a neural networks are adapted through a process of stimulation by the environment in which the network is embedded.

Classification of Learning Algorithms

Supervised Learning

Every input pattern that is used to train the network is associated with an output pattern. A teacher is assumed to be present during the learning process, when a comparison is made between the networks computed output and correct expected output to determine the error. The aim is to minimize the operation error of the network. Perceptron learning rule and Back propagation algorithm follows supervised learning.

Unsupervised Learning

In this learning method, the target output is not presented to the network. It is, as if there is no teacher to present the desired patterns and hence, the system learns of its own by discovering and adapting to structural features in the input patterns. Such learning can be used for classification tasks.

Typical algorithms include:

- Hebbian rule
- Competitive learning
- Kohonen algorithm.
- Self-Organizing Map (SOM)
- Adaptive resonance theory (ART)

Reinforced learning

In this method, a teacher though available, does not present the expected answer but only indicates if the computed output is correct or incorrect. The information provided helps the networks in its learning process. A reward is given to the correct answer computed and penalty for wrong answer. But, reinforced learning is not one of the popular forms of learning.

Supervised and Unsupervised learning methods have various rules. They are

1. Hebbian learning

This rule was proposed by hebb (1949) and is based on correlative weight adjustment. This is the oldest learning mechansim. In this, the input-output patterns pairs (X_i, Y_i) are associated by the weight matrix W, known as the correlation matrix. It is computed as

$$W = \sum_{i=1}^{n} X_i Y_i^T$$

Here Y_i^T is the transpose of the associated output vector Y_i, Numerous variants of the rule have been proposed.

2. Gradient descent learning

This is based on the minimization of error E defined in terms of the weights and activation function of the network. Also it is required that the activation function employed by the network is differentiable, as the weight update is dependent on the gradient of the error E.

3. Competitive learning

In this method, those neurons which represent strongly to input stimuli have their weights updated. When an input pattern is present, all neurons in the layer compete and the wining neuron undergoes weight adjustment. Hence it is a "winner-takes-all" strategy.

4. Stochastic learning

The weights are adjusted in probabilistic fashion.

Artificial Neural Networks in Agriculture

The Applications of Artificial Neural Networks in Agriculture forms the basis for

- Grain Inspector
- Taming the Tamarisk
- Predict outcome of future events

Grain Inspector

The little black box for grain inspectors : A Canadian-made device could soon take the subjectivity and guesswork out of grading the quality of wheat. "For generations, grading grain has followed the same basic script: An inspector takes a sample of grain, looks it over, mentally compares it to all the grain he has seen before, and gives it a grade. Agricultural Canadian companies have developed a new device that uses digital imaging and artificial neural networks to visually analyze and grade grain. Grain inspector picks up a handful of grain, looks for signs of mildew, frost and other damage, and grades it on a scale of one to four. Digital imaging and artificial intelligence helps to analyze and grade grain on a much more detailed level."

Tamming the Tamarisk

Taming the Tamarisk is a new Geographic Information System (GIS) weapon helps a county fight a dangerous shrub. "Growing from 5 to 20 feet tall, the tamarisk [also known as the salt cedar] was originally introduced into the United States from Eurasia in the early 1800's as an ornamental plant. Because of its dense, deep root system, settlers in South-western states planted salt cedar along streambeds to prevent erosion from flooding. Over the years, tamarisk spread from the Colorado River basin to New Mexico, and as far east as Texas. Feature Analyst from Visual Learning Systems uses machine learning technology to classify object-specific geographic features specified by the user. They actually go into the field with a sub meter GPS unit, and gather training samples and accuracy assessment samples for salt cedar. Using these, they train the computer to recognize salt cedar as opposed to other types of vegetation. And then finally, running those algorithms, it will spit out the area that can be used in most GIS applications. Without these enhancement stages, salt cedar could not be pulled out.

Predict outcome of future events

The neural network process develops a model that can be used to predict outcomes of future events. Also, a model for Crop data was developed that predicted how 15 variables - including rainfall, temperature, soil moisture, and fertilizer - influenced crop yields. The analysis of the data revealed some insights that previous data analyses had not. The model shows the rainfall amounts especially rainfall during the long, hot July days. Rainfall was more important than amounts or types of fertilizers used, crop rotation cycles, or nitrogen levels in the soil. Such knowledge discovery techniques lead to accurate predictions. The ability to accurately predict outcomes can change how farmers practice their trade.

The input layer consists of measured variables or inputs fed into input nodes. Various weights are attached to the inputs that determine how the inputs interact, and the sum of the inputs passes through a hidden layer where the network performs problem specific sub functions and reaches an output value. This output is compared to a known outcome, and the process is repeated using new weights in an effort to get closer to the actual outcome. The end result of a neural network is an accurate predictive model.

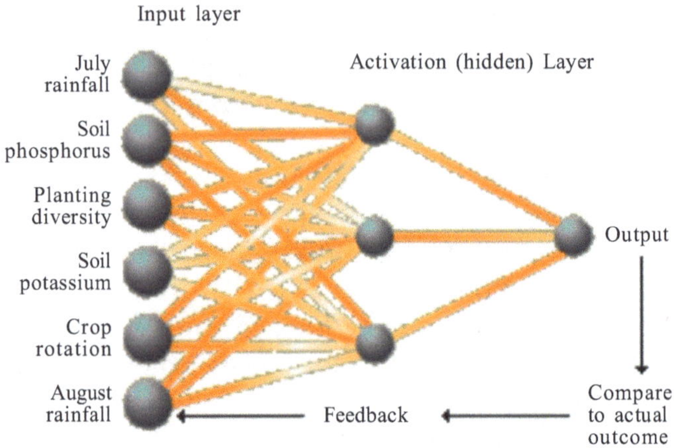

Neural network

A neural network is a nonlinear function that maps multiple input variables to possible outputs in a way that is similar to a real nervous system. Whereas a nervous system has neurons that take in sensory data, a neural network has inputs that pass through network nodes. Instead of axons and dendrites carrying messages between nerve cells, a neural network has weights and paths that determine how inputs interact and influence the final output. A multilayered neural network starts with an input layer and ends with the output layer. In between are one or more hidden layers that allow the neural network to construct problem-specific sub functions needed to solve difficult problems.

Every node in a neural network sums up its inputs and passes the sum through the nonlinear activation function to reach an output value. Each node passes its information in succession until the output layer and final predictions are reached.

A neural network is "trained" to reach an accurate prediction. The process begins by scrambling all the data so that nothing remains but a random mapping among inputs and outputs. Starting with this blank slate, researchers feed known input and output variables into the system. The input variables start the network process and an outcome is reached and compared to the known or actual outcome. Information from this comparison is then fed back through the network, and weights are changed gradually to bring the network's outcome closer to the actual outcome. If all goes well, the network becomes more and more accurate each time it adjusts. Eventually, researchers have an accurate model that can be applied to future problems.

Advantages

- A neural network can perform tasks that a linear program cannot.
- When an element of the neural network fails, it can continue without any problem by their parallel nature.
- A neural network learns and does not need to be reprogrammed.
- It can be implemented in any application.

Other advantages include:

1. *Adaptive learning*: An ability to learn how to do tasks based on the data given for training or initial experience.
2. *Self-Organization*: An Artificial Neural Network can create its own organization or representation of the information it receives during learning time.
3. *Real Time Operation*: Artificial Neural Network computations may be carried out in parallel and special hardware devices are being designed and manufactured which take advantage of this capability.
4. *Fault Tolerance via Redundant Information Coding*: Partial destruction of a network leads to the corresponding degradation of performance. However, some network capabilities may be retained even with major network damage.

Disadvantages

- The neural network needs training to operate.
- Requires high processing time for large neural networks.

Intelligent Agent

An intelligent agent is software that assists people and acts on their behalf. Intelligent agents work by allowing people to delegate work that they could have done to the software agent. Agents can just as assistants, can automate repetitive tasks, remember things we forgot, intelligently summarize complex data, learn from us and make recommendations to us.

As the above definition suggests, intelligent agents are software programs designed to perform the same kind of tasks that human agents do. Human agents have been around for thousands of years and include travel agents, insurance agents, and literary agents and may others. A few of the tasks commonly performed by agents include providing information about a service or product, locating the best sources for any goods are information required, finding the best price for an item, negotiating agreements, and monitoring results and resolving problems. Most of these tasks can now be performed by smart software applications known as intelligent agents or software robots.

These applications are said to be 'intelligent' because users can give an agent a relatively high-level goal to achieve and it will then figure out the steps necessary to accomplish the goal, rather than the user having to give explicit instructions to the program for each step of the process. In addition, agents are also usually able to learn the specific needs and interests of their users and adapt to them over time.

How are agents used for?

Intelligent agents and software robots are increasingly being used to deal with the vast amount of information available on the internet and the increasing complexity of computer systems and software. Such agents are frequently used to find or filter information for a user, identify patterns and trends from very large amounts of data and act as interactive assistants for computer interfaces.

Agents are also used most commonly on the web in search engines and the e-commerce capacities, but they also have important applications for the libraries, information filtering and management, and personal use.

Classes of Intelligent Agents

Agents are grouped into five classes based on their degree of perceived intelligence and capability as follows:

1. simple reflex agents
2. model-based reflex agents
3. goal-based agents
4. utility-based agents
5. learning agents

1. Simple reflex agents

A simple reflex agent acts only on the basis of the current percept. The agent function is based on the condition-action rule: if condition then action. This agent function only succeeds when the environment is fully observable. Some reflex agents can also contain information on their current state which allows them to disregard conditions whose actuators are already triggered.

2. Model-based reflex agents

Model-based agents can handle partially observable environments. Its current state is stored inside the agent maintaining some kind of structure which describes the part of the world which cannot be seen. This behavior requires information on how the world behaves and works. This additional information completes the "World View" model. A model-based reflex agent keeps track of the current state of the world using an internal model. It then chooses an action in the same way as the reflex agent.

3. Goal-based agents

Goal-based agents are model-based agents which store information regarding situations that are desirable. This allows the agent a way to choose among multiple possibilities, selecting the one which reaches a goal state.

4. Utility-based agents

Goal-based agents only distinguish between goal states and non-goal states. It is possible to define a measure of how desirable a particular state is. This measure can be obtained through the use of a utility function which maps a state to a measure of the utility of the state.

5. Learning agents

Learning has an advantage that it allows the agents to initially operate in unknown environments and to become more competent than its initial knowledge alone might allow.

Other Classes of Intelligent Agents

Intelligent Agent or a complete Intelligent Agent are:

- Decision Agents (that are geared to decision making)
- Input Agents (that process and make sense of sensor inputs - e.g. neural network based agents)

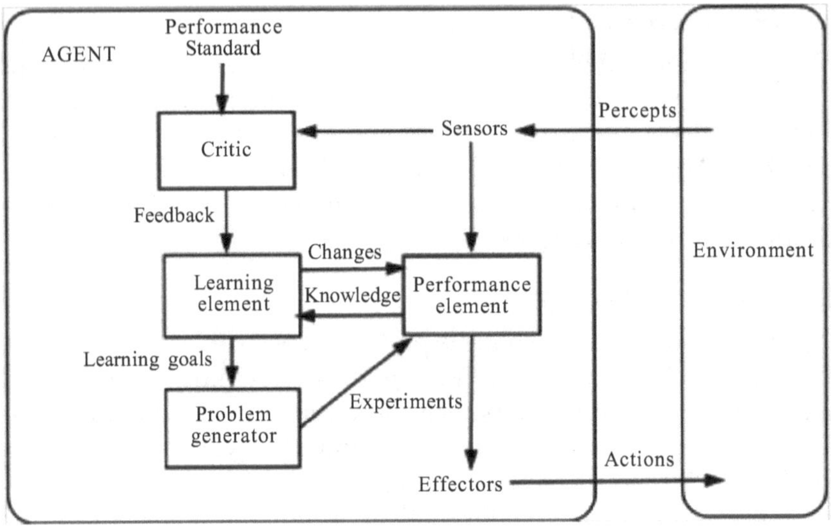

Learning Agent

- Processing Agents (that solve a problem like speech recognition)
- Spatial Agents (that relate to the physical real-world)
- World Agents (that incorporate a combination of all the other classes of agents to allow autonomous behaviors)
- Believable agents - An agent exhibiting a personality via the use of an artificial character (the agent is embedded) for the interaction.
- Physical Agents - A physical agent is an entity which percepts through sensors and acts through actuators.

Temporal Agents - A temporal agent may use time based stored information to offer instructions or data acts to a computer program or human being and takes program inputs percepts to adjust its next behaviors.

Characteristics of Intelligent Agents

A software application is considered to be intelligent if at all it has certain characteristics.

It is commonly agreed that true intelligent agents must prove themselves to be:

Autonomous: able to exercise control over their own actions.

Proactive: be goal oriented and able to accomplish goals without performing from a user, and recognize and react to changes in their environment.

Persistent: keep running a process continually until the desired result has been achieved

Reasoning: be able to reason out their actins.

Productive: capable of achieving desired results

Communicative: be able to communicate both with humans and other agents.

Adaptable: able to function on multiple platforms and solve technical difficulties on their own

Mobile: able to roam networks freely

Very few intelligent agents in use today fulfill all of the criteria for 'true intelligence'. In particular, today's agents have difficulty in communicating with one another and in being able to function on multiple platforms, due to the lack of standards for inter-agent communication. However, a software application must demonstrate at the very least autonomous decision-making abilities to merit the name 'agent'.

Exercise to Practice

1. What is a Neural Network?
2. How does an artificial neural network works?
3. Explain the characteristics of neural networks.
4. Write notes on architectures of artificial neural network.
5. Define
   Supervised learning
   Unsupervised learning
   Intelligent agent
6. Write in detail about supervised and unsupervised learning method.
7. Application of artificial neural networks in agriculture.
8. Advantages of Artificial Neural Networks

CHAPTER 8

Expert Systems

Introduction

An Expert System (ES), also called a Knowledge Based System (KBS), is a computer program designed to simulate the problem-solving behavior of an expert. An expert system is a computer application that solves complicated problems that would otherwise require extensive human expertise. To do so, it simulates the human reasoning process by applying specific knowledge and interfaces. Expert systems also use human knowledge to solve problems that normally would require human intelligence. These expert systems represent the expertise knowledge as data or rules within the computer. These rules and data can be called upon when needed to solve problems.

A system that uses human knowledge captured in a computer to solve problems that ordinarily require human expertise. A computer program designed to model the problem solving ability of a human expert. An intelligent computer program that uses knowledge and inference procedures to solve problems that was difficult enough to acquire significant human expertise for their solutions.

Expert systems typically have a number of several components.

- Knowledge base is the component that contains the knowledge obtained from the domain expert. Normally the way of representing knowledge is using rules.
- Inference engine is the component that manipulates the knowledge found in the knowledge base as needed to arrive at a result or solution.
- User interface is the component that allows the user to query the system and receive the results of those queries.

What is Meant by the Term Expert?

- Specialized skill
- Experience
- Knowledge
- Tricks - Shortcuts to resolve a problem efficiently.

What can Expert Systems can do?

In agriculture, expert systems are capable of integrating the perspectives of individual disciplines (e.g. plant pathology, entomology, horticulture, agricultural meteorology) into a framework that best addresses the type of decision-making required for modern farmers. Expert systems can be one of the most useful tools for accomplishing the task of providing growers with the day-to-day, integrated decision support needed to grow their crops.

Building Blocks of Expert Systems

Expert systems are composed of several components such as a user interface, a database, a knowledge base and an interface mechanism. Moreover, Expert system development usually proceeds through several phases including problem selection, knowledge acquisition, knowledge representation, programming, testing and evaluation.

User Interface

The function of the user interface is to present questions and information to the user and supply the user's responses to the inference engine. The questions are mostly in the form visuals that are developed as images, animation clips and video clips. Any values entered by the user must be received and interpreted by the user interface. The user interface checks all responses to ensure that they are of the correct data type. Whenever the user enters an illegal response, the user interface informs the user that his input was invalid and prompts him to correct it.

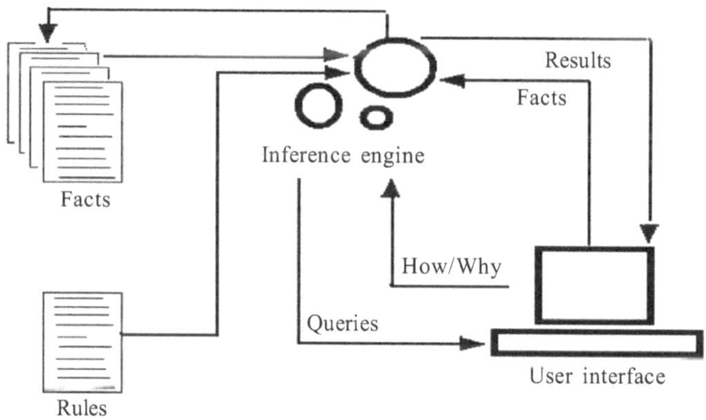

Components of Expert systems

Knowledge Base

The knowledge the expert uses to solve a problem must be represented in a fashion that can be used to code into the computer and then be available for decision making by the expert system. Every expert system consists of two principal parts: the knowledge base and the reasoning, or inference, engine.

The knowledge base of expert systems contains both factual and heuristic knowledge. Factual knowledge is the knowledge of task domain that is widely shared, typically found in textbooks or journals, and commonly agreed upon by those knowledgeable in the particular field.

Heuristic knowledge is less rigorous, more experiential, more judgmental knowledge of performance. In contrast to factual knowledge, heuristic knowledge is largely individualistic. It is the knowledge of good practice, good judgment and plausible reasoning in the field. It is the knowledge that underlies the art of good guessing.

Knowledge representation formalizes and organizes the knowledge. One widely used representation is the production rule, or simply rule. A rule consists of an IF part and a THEN part (also called a condition and an action).

- The IF part lists a set of conditions in some logical combination. The piece of knowledge represented by the production rule is relevant to the line of reasoning being developed if the IF part of the rule is satisfied.
- The then part can be concluded, or its problem-solving action taken.

Expert systems whose knowledge is represented in rule form are called rule-based systems. Another widely used representation, called the unit (also known as frame, schema, or list structure) is based upon a more passive view of knowledge. The unit is an assemblage of associated symbolic knowledge about an entity to be represented. Typically, a unit consists of a list of properties of the entity and associated values for those properties.

The problem-solving model, or paradigm, organizes and controls the steps taken to solve the problem. One common but powerful paradigm involves chaining of IF-THEN rules to form a line of reasoning.

- *Forward chaining*: If the chaining starts from a set of conditions and moves toward some conclusion, the method is called forward chaining.
- *Backward chaining*: If the conclusion is known (for example, a goal to be achieved) but the path to that conclusion is not known, then reasoning backwards is called for, and the method is backward chaining.

These problem-solving methods are built into program modules called inference engines or inference procedures that manipulate and use knowledge in the knowledge base to form a line of reasoning.

Reasoning of Uncertainty

- Knowledge is always and almost incomplete and uncertain.
- To deal with uncertainty, a rule may have associated with a confidence factor or a weight.
- An important subclass of methods of reasoning with uncertainty is called fuzzy logic used in fuzzy systems.

Inference Mechanism

The inference mechanism will be integrated as a software program (inference engine) that the part of the program containing reasoning capability. It interacts with the knowledge base (IF...THEN...ELSE) which contains the information about how to solve problems within the problem domain.

This is the global memory where the knowledge based system records information relating to a specific problem that is trying to solve. Much of this information comes from the user but the memory is also used by the inference engine to record its own conclusions and to remember its chain of reasoning. By comparing what it knows about the problem domain general with what it knows about the specific problem, the inference engine tries to proceed logically towards a better solution.

Inference Rule

An understanding of the "inference rule" concept is important to understand expert systems. An inference rule is a statement that has two parts, an if-clause and a then-clause. This rule is what gives expert systems the ability to find solutions to diagnostic and prescriptive problems. An example of an inference rule is:

If the symptom of crop is X, Then the nutrition deficiency is Y. An expert system's rule base is made up of many such inference rules. They are entered as spate rules and it is the inference engine that uses them together to draw conclusions. One advantage of inference rules over the traditional programming is that inference rules use reasoning, which more closely resembles human reasoning. Thus, when a conclusion is drawn, it is possible to understand how this conclusion was reached. Furthermore, because the expert system uses knowledge in a form similar to the human expert, it may be easier to retrieve this information from the expert.

Designing Expert Systems in Problem Solving

The architecture of expert systems consists of the following steps:

1. The sequence of steps taken to reach a conclusion is dynamically synthesized with each new case. It is not explicitly programmed when the system is built.

2. Expert systems can process multiple values for any problem parameter. This permits more than one line of reasoning to be pursued and the results of incomplete reasoning to be presented.
3. Problem solving is accomplished by applying specific knowledge rather than specific technique. This is a key idea in expert systems technology. It reflects the belief that human experts do not process their knowledge differently from others, but they do possess different knowledge. With this philosophy, when one finds that their expert system does not produce the desired results, work begins to expand the knowledge base, not to re-program the procedures.

There are various expert systems in which a rule base and an inference engine cooperate to simulate the reasoning process that a human expert pursues in analyzing a problem and arriving at a conclusion. In these systems, in order to simulate the human reasoning process, a vast amount of knowledge needed to be stored in the knowledge base. Generally, the knowledge base of such an expert system consist of a relatively large number of "if then" type of statements that were interrelated in a manner that, in theory at least, resembled the sequence of mental steps that were involved in the human reasoning process.

Most expert systems, in the past, have been run only on large information handling systems. Recently, the capacity of personal computers has increased to a point where it is becoming possible to consider running some types of simple expert systems on them.

In some applications of expert systems, the nature of the application and the amount of stored information necessary to simulate the human reasoning process for that application is just too vast to store in the active memory of a computer. In other applications of expert systems, the nature of the application is such that not all of the information is always needed in the reasoning process.

An example of the latter type application would be the use of an expert system to diagnose a data processing system comprising many separate components, some of which are optional. When that type of expert system employs a single integrated rule base to diagnose the minimum system configuration of the data processing system, much of the rule base is not required since many of the components which are optional units of the system will not be present in the system.

The user interface and the procedure interface are the two important functions in the information collection process.

End User

The end-user sees an expert system through an interactive dialog, an example of which follows:

Q: Do you know which restaurant you want to go to? A: No

Q: Is there any kind of food you would particularly like? A: No

Q: Do you like spicy food? A: No

Q: Do you usually take soft drinks with meals? A: Yes

As can be seen from this dialog, the system is leading the user through a set of questions, the purpose of which is to determine a suitable set of restaurants to recommend. This dialog begins with the system asking if the user already knows the restaurant choice (a common feature of expert systems) and immediately illustrates a characteristic of expert systems, users may choose not to respond to any question. In expert system, dialogs are not pre-planned. There is no fixed control structure. Dialogs are synthesized from the current information and the contents of the knowledge base. Because of this, not being able to supply the answer to a particular question does not stop the consultation.

During the consultation, the rule base is searched for conditions that can be satisfied by facts supplied by the user. This operation is performed by the inference engine. Once all of the conditions (i.e. IF parts of rules) of a rule are matched, the rule is executed and the appropriate conclusion is withdrawn. Based upon the conclusions drawn and the facts obtained during consultation, the inference mechanism determines which question will be asked and in what order. There are various inference methods available to perform the tasks of searching, matching and execution.

A distinctive characteristic of expert systems that distinguishes them from conventional programs is their ability to utilize incomplete or incorrect data. Given only a partial data set, an expert is likely to have less than absolute certainty in the conclusion. The degree of certainty can be quantified in relative terms and included in the knowledge base. The certainty values are assigned by the expert during the knowledge acquisition phase of developing the system. By incorporating rules in the knowledge base with different certainty values, the system will be able to offer solutions to problems without a complete set of data. The capacity to deal with uncertainty is available in development software.

Applications of Expert System

There are several major application areas of expert system such as agriculture, education, environment, law manufacturing, medicine power systems etc. Expert systems combine the experimental and experiential knowledge with the intuitive reasoning skills of a multitude of specialists to aid farmers in making the best decisions for their crops. Expert systems are used in applications such as medical diagnosis, equipment repair, investment analysis, financial, estate and insurance planning, route scheduling for delivery vehicles, contract bidding, counseling for self-service customers, production control and training.

Expert systems are the computer applications that perform a task that would otherwise be performed by a human expert.

- Programming computers to make decisions in real-life situations.
- Examples
  - diagnose human illnesses
  - make financial forecasts
  - schedule routes for delivery vehicles

The early state of developing the Expert Systems are in the 1960s and 1970s were typically written on a mainframe computer in the programming language based on list Processing (LISP). A general-purpose tool for developing expert systems is called a "shell".

An Expert System for Decision Problem

The expert system can be used as a stand alone advisory system for the specific knowledge domain perhaps with monitoring by a human expert. It also can provide decision support for a high-level human expert. The agriculture expert system also allows a high-level expert to be replaced by a subordinate expert aided by the expert system.

The main purposes of the expert system are:

- Delivery system for extension information,
- Providing management education for decision makers (farmers), and

- Dissemination of up-to-date scientific information in a readily accessible and easily understood form to agricultural researchers, advisers and farmers.
- By the help of the expert system, the farmers can produce a more high quality product to the citizen.

Expert Systems are Used to Aid

1. Single point decisions. e.g: Planning
2. Designing. e.g: Design of an irrigation system
3. Selection. e.g: The most suitable Crop variety or market outlet
4. Diagnosis or identification. e.g: a livestock disorder
5. Interpretation. e.g: a set of financial accounts
6. Prediction. e.g: Extreme events such as thunderstorms and frost.

Advantages of Expert System

There are few advantages that the expert system being use in the agriculture field. It has the ability to imitate human thought and reasoning.

1. The expert system makes modification of knowledge very convenient.
2. It has the ability of interpretation and transparence makes interaction and more user friendly.
3. With the machine learning technique knowledge can be acquired automatically and directly from experimental data. Real time examples help to provide the right information which is timely and actionable.
4. It can provide expert level recommendations understandable to users (farmers).
5. It can hold and maintain significant levels of information.
6. It has the ability to handle uncertain information.
7. It never 'forgets' to ask a question, as a human right.

Disadvantages of Expert System

- Lacks common sense needed in some decision making.
- Cannot make creative responses as human expert would in unusual circumstances.

- Domain experts not always able to explain their logic and reasoning.
- Cannot adopt to changing environment, unless knowledge base is changed.

Cases of Expert Systems in Agriculture

Rice-Crop Doctor

National Institute of Agricultural Extension Management (MANAGE) has developed an expert system to diagnose pests and diseases for rice crop and suggest preventive/curative measures. The rice crop doctor illustrates the use of expert-systems broadly in the area of agriculture and more specifically in the area of rice production through development of a prototype, taking into consideration a few major pests and diseases and some deficiency problems limiting rice yield.

The following diseases and pests have been included in the system for identification and suggesting preventive and curative measures. The diseases included are rice blast, brown spots, sheath blight, rice tungro virus, false smut fungi, bacterial leaf blight, sheath rot and zinc deficiency disease. The pests included are stem borers, rice gall midge, brown plant hopper, rice leaf folder, green leaf hopper and Gundhi bug.

AGREX

Center for Informatics Research and Advancement, Kerala has prepared an Expert System called AGREX to help the Agricultural field personnel give timely and correct advice to the farmers. These Expert Systems find extensive use in the areas of fertilizer application, crop protection, irrigation scheduling, and diagnosis of diseases in paddy and post harvest technology of fruits and vegetables.

Distributed Knowledge Based Expert System (KBES) in Agriculture

There are three levels in which the basic process of agriculture is happening.

1. Low level farmers.
2. Middle level agricultural officers.
3. Higher level research institutions.

Each level requires a data depending upon its requirements with interrelationships among them. The farmers interface is designed in such a way that the communication between the expert system and the farmer will be in the farmers own native language. The knowledge base acts as a bridge between farmers and research institutions. The production systems at farmers' level knowledge base which are not able to find an answer are formed as unanswered dynamic framers and tried for solution by agricultural officers.

These frames are transferred to research institutions with the same additional view of agricultural officers. Hence research institutions can come to know about the new undiscovered problems that exist at farmers' level.

Expert Systems for Cotton Crop Management

This Expert System has been developed by the US Department of Agriculture to provide appropriate management recommendations to cotton growers.

CALEX

This is a blackboard based integrated expert decision support system for agricultural management, developed at University of California. CALEX can be used by growers, pest control advisors, consultants and other managers.

VARIEX

This expert system developed at Technical University of Brno, Czechoslovakia enables selection of the best cultivators for different agricultural situations.

Crop Growth Control Expert System

There are fertilizing, irrigating, spraying insecticide process and adopting other measures in crop management, which rely on crop state analysis. Cotton Crop Management Expert System provides support for crop growth control system (CGCS), i.e., advises growers about optimal population and structure of crop in planting stage and when and what to adopt measures on their crop to keep at optimal state, to avoid Infestation in various stage of growth and development of crop and finally to obtain the highest crop productivity in harvest stage

CLIPS Expert System

A prototype management CLIPS (C Language Integrated Production System) expert system has been developed by agricultural scientists. This form of artificial intelligence provides an extension tool, which will enable farmers to very economically use a computer program to reach conclusions concerning profitable production that normally would require consultation with a forage expert.

To date, the management considerations and the sequence in which they were built into the knowledge base are: soil drainage, soil pH, soil P test, soil K test, and chemical weed control, and expected longevity of stand, variety recommendation, method and rate of seeding and pure live seed. The knowledge base of these CLIPS expert system can easily be updated, as new information becomes available or can be modified for use in other states or farming regions.

DSS4Ag

The Decision Support System for Agriculture (DSS4Ag) is an expert system being developed by the Site-Specific Technologies for Agriculture (SST4Ag) precision farming research project at the INEEL. DSS4Ag uses artificial intelligence and computer science technologies to make spatially variable, site-specific, economically optimum decisions on fertilizer use.

The DSS4Ag has an open architecture that allows for external input and addition of new requirements and integrates its results with existing agricultural systems' infrastructures. The DSS4Ag reflects a paradigm shift in the information revolution in agriculture that is precision farming.

The Expert system must be developed in local languages which will help the Farmers to develop their own expertise which in turn will enhance the production and productivity of Agriculture. Expert systems must be available in village booths which act as information resource center for the farmers in the villages.

Prototype Decision Support System

A prototype decision support system is being used to assist managers with the development, implementation and management of frost protection. The program is divided into four main modules like

1. Frost protection strategies
2. Operational management
3. Forecasting (and)
4. Instrumentations

Satellite Downlink Frequency

Using a unique downlink facility, real time meteorological data are being processed on a network of advanced computer workstations. This data feeds a series of computer models that generate site specific predictions of temperature, wet bulb temperature, and other parameters used to alert fruit, vegetable and nursery growers. Model outputs are distributed to Extension specialists and others over a campus-wide fiber optic communications system and to all country Extension offices over the ACENET (Applied Catalysis European NETwork) which tries to bring coherence and foster cooperation between national research programmes and policies on applied catalysis.

Its activities have:

- Integrated processes between the national research-funding organizations, as a result of sharing good practice and infrastructure.
- Increased efficiency in research by promoting interdisciplinary knowledge-sharing.
- Provided a framework for the education and qualification of young scientists following an inventory of real and genuine training needs.

During critical freeze situations, information can be disseminated live via satellite to media and rural satellite dish owners using the uplink facility. A pilot network of automated weather stations incorporating a 10-m tower for monitoring inversions has been installed to enhance the program.

Exercise to Practice

1. Explain the tasks involved in building expert system?
2. Application of expert system in agriculture.
3. Advantage and disadvantage of expert system

4. Write notes on distributed knowledge based expert system in agriculture.
5. Explain about Rice Crop Doctor.

CHAPTER 9

Decision Support System (DSS)

Introduction

Decision support systems constitute a class of computer-based information systems including knowledge-based systems that support decision-making activities.

A properly-designed DSS is an interactive software-based system intended to help decision makers compile useful information from raw data, documents, personal knowledge and/or business models to identify and solve problems and make decisions.

Three fundamental components of a DSS architecture are

1. The database (or knowledge base),
2. The model (i.e., the decision context and user criteria) and
3. The user interface.

DSS technology levels (of hardware and software) may include:

- The actual application that will be used by the user. This is the part of the application that allows the decision maker to make decisions in a particular problem area. The user can act upon that particular problem.

- Generator contains Hardware/software environment that allows people to easily develop specific DSS applications. This level makes use of case tools or systems such as Crystal.
- Tools include hardware/software.

DSS components may be classified as:

- *Inputs*: Factors, numbers and characteristics to analyze.
- *User Knowledge and Expertise*: Inputs requiring manual analysis by the user.
- *Outputs*: Transformed data from which DSS "decisions" are generated
- *Decisions:* Results generated by the DSS based on user criteria

Types of DSS

Communication-driven Decision Support System

A communication-driven Decision Support System supports more than one person working on a shared task. A Communications Driven DSS models is when many collaborators work together to come up with a series of decisions to set in motion a solution or strategy. This communications driven DSS model can be in an office environment or on the web.

It is a type of DSS that enhances decision-making by enabling communication and sharing of information between groups of people. At its most basic level a C-D DSS could be a simple threaded e-mail. At its most complex, it could be a web-conferencing application or interactive video.

Communication-Driven DSS will exhibit at least one of the following characteristics:

- Supports coordination and collaboration between two or more people;
- Facilitates information sharing;
- Enables communication between groups of people;
- Supports group decisions;

Data-Driven Decision Support System

A data-driven Decision Support System or data-oriented DSS emphasizes access to and manipulation of a time series of internal company data and sometimes, external data. A Data Driven DSS model puts its emphasis on collected data that is then manipulated to fit the decision maker's needs. This data can be internal, external and in a variety of formats. It is important that usually data is collected and categorized as a time series which is a collection of data that forms a sequence such as daily sales, operating budgets from one quarter to the next, inventory levels over the previous year, etc.

Data-driven DSS are a form of support system that focuses on the provision of internal (and sometimes external) data to aid decision making. Most often this will come in the form of a data warehouse - a database designed to store data in such a way as to allow for its querying and analysis by users.

Another example of a data-driven DSS would be a Geographic Information System (GIS), which can be used to visually represent geographically dependant data using maps.

Document-Driven Decision Support System

A document-driven Decision Support System manages, retrieves, and manipulates unstructured information in a variety of electronic formats. A Document Driven DSS model uses documents in a variety of data types such a text documents, spreadsheets and database records to come up with decisions as well as further manipulate the information to refine strategies.

Document-driven DSS are support systems designed to convert documents into valuable business data. While data-driven DSS rely on data that is already in a standardized format that lends itself to database storage and analysis, document-driven DSS makes use of data that cannot easily be standardized and stored.

The three primary forms of data used in document driven DSS are:

- Oral (i.e. transcribed conversations);
- Written (i.e. reports, memos, e-mail and other correspondence);
- Video (i.e. TV commercials and news reports).

None of these formats lend themselves easily to standardized database storage and analysis, so managers require DSS tools to convert them into data that can be valuable in the decision making process.

Document-driven DSS is the newest field of study in Decision Support Systems. Examples of document-driven tools can be found in Internet search engines, designed to sift through vast volumes of unsorted data through the use of keyword searches.

Knowledge-Driven Decision Support System

A knowledge-driven Decision Support System provides specialized problem-solving expertise stored as facts, rules, procedures or in similar structure. A Knowledge Driven DSS model uses special rules stored in a computer or used by a human to determine whether a decision should be made. For instance, for many day traders a stop loss limit can be seen as a knowledge driven DSS model. These rules or facts are used in order to make a decision. Knowledge-driven DSS are systems designed to recommend actions to users. Typically knowledge-driven systems are designed to sift through large volumes of data, identify hidden patterns in that data and present recommendations based on those patterns.

Model-Driven Decision Support System

A model-driven Decision Support System emphasizes access to and manipulation of a statistical, financial, optimization or simulation model. Model-driven DSS use data and parameters provided by users to assist decision makers in analyzing a situation; they are not necessarily data-intensive.

Model-based decision support systems can be extremely useful in forecasting the effects of changes in business processes, as they can use past data to answer complex 'what-if' questions for decision makers.

In addition to these basic types of DSS there are also two additional factors:

- Spreadsheet Decision Support System,
- Web-based Decision Support System.

Spread-sheet Decision Support System

Model and Data driven DS systems can be built using spreadsheets. Spreadsheets offer decision-makers easy to understand representations of large amounts of data. Additionally, spreadsheet data is arranged in such a way as to make it easy to convert the data into visualizations to further aid decision-makers.

Web-based Decision Support System

Any type of DSS can be web-based. The term simply describes any decision support system that is operated through the interface of a web browser, even if the data used for decision support remains confined to a legacy system such as a data warehouse.

Scope of DSS

In addition to these basic types of Decision Support System there are also two separate categories used to define systems.

1. Enterprise-wide Decision Support System

Enterprise-wide DS systems are systems that are linked into large data warehouses and offer decision support to managers at all levels of an enterprise. Enterprise-wide systems will typically be basic, general use systems that can perform a wide variety of functions.

2. Desktop Decision Support System

Desktop DS systems are much smaller applications designed to be run from a desktop PC. While these systems may well be linked into a data warehouse or other large volume of data, they will typically be more limited in scope. An example of a desktop DSS is Microsoft Excel, the desktop spreadsheet application.

Benefits of DSS

- Improves personal efficiency
- Expedites problem solving (speed up the progress of problems solving in an organization)
- Promotes learning or training
- Increases organizational control
- Generates new evidence in support of a decision
- Reveals new approaches to thinking about the problem space
- Helps automate the managerial processes

Common Uses of Decision Support Systems

- Application of Fertilizers - type and amount
- Use of Pesticides - deciding the proper threshold
- Deciding the cultivar
- Deciding the suitable date for harvest
- Optimizing agricultural benefits - important for farmers

An example decision: Pesticide Spraying

Should the field be sprayed? - is there sufficient risk of pests (insects, diseases or weeds)

What pesticide to use? - Differences in price, effectiveness and regulations.

When should the field be sprayed? - Weather, especially rain, greatly affects spraying

Decision Support System for Farm Management

In India, about 70% of population depends upon agriculture. The credit of the increased production of the agriculture products in the past could be given to the efforts of farmers. When the production is stagnating it has become essential that the farmers collect important and updated information about any of the crop. Keeping this in view, there is a need for development of dynamic website for farm entrepreneur which could help them in day-to-day decision taking. The business of farming has entered a new era - an age where key to success is perfect, timely information and careful decision-making. The farmer has to decide about various production options utilizing the results of latest developments of research and technology. Informed and quick decision making is therefore required to ensure profitable performance of the farmers.

Decision support systems are a class of computer-based information systems including knowledge based systems that support decision making activities. DSS is a computerized system to help in making decisions. A decision is a choice between alternatives based on estimates of the values of those alternatives. Supporting a decision means helping people working alone or in a group gathers intelligence, generate alternatives and make choices. Supporting the choice making process involves supporting the estimation and evaluation.

In practice, references to DSS are usually references to computer applications that perform such a supporting role. DSS are "interactive computer-based systems that help decision makers utilize data and models to solve unstructured problems".

The decision process may be divided into three phases: intelligence, choice and design as shown below:

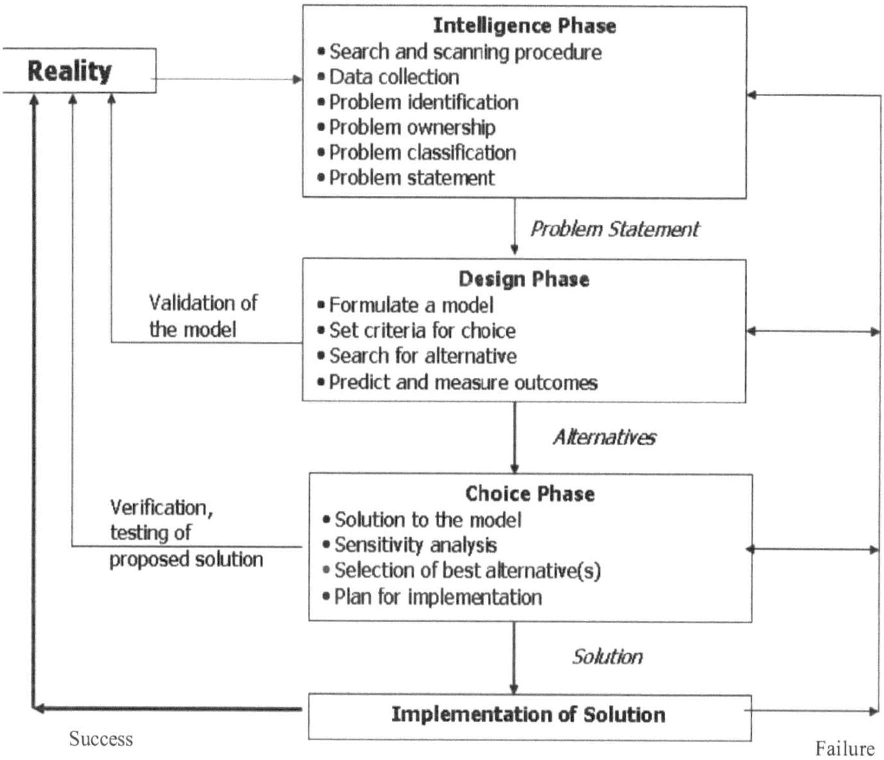

Phases of Decision Support System

The timely information to the farmers is closely linked to the Agricultural Development and well being of the rural communities. Quick information transfer between the researchers and the farmers has specific importance. Hence there is a focus on the problem of developing a decision support system for farm entrepreneur.

The farmers will be able to access the necessary information regarding the various government policies for agriculture, the ongoing socio-economic research and the different benefits for the entrepreneur for investing in particular area of farm management.

The Farmers can acquire this information from:

- Farmer's organizations
- Researchers
- Other Farmers
- Farm journals, radio, television, Government Departments and agencies.

The basic fundamental of any project involving the database is the collection of data and information. This project also requires the collection of information from various sources.

The main source of the data is the service published by the state or center Government. This will help the farmers to take interactive, flexible and quick decisions. This will also help them in increasing their productivity by raising yield per hectare in food grain, thus leading to their economic growth. It will keep track of farmer's all type of information related to crops.

DSS with all the ready information, help the farmers in a very useful manner. The farmers can get all the information just at click of the mouse and they need not to travel to Agricultural Universities for that.

The main advantage of this model to the farmer is that they can retrieve the dynamic information for their farm management decisions. DSS Framework being an agent for the driving force behind the changes in highland resource uses, the farm or household is considered to be the center of this analysis.

The decisions on agricultural land and water uses are made in response to resource endowments, economic conditions and socio-cultural norms of the household or communities. The Decision Support System, can aid assessing management options to help resolving or avoiding land and water use conflicts.

The model will help the farmers in increasing their productivity by raising the yield/hectare in food grains: thus, leading to their economic growth. This system has been developed to keep track of farmers all type of information related to crops.

Certain applications that are successfully developed using this database are:

- Farmers can manage their cash flow through the DSS system in a more predictable and efficient way. It is a more common problem with the farmers to manage the cash received at time of harvesting the crop.
- They can avail full benefit of their cash management by co-relating it with the loans and advances.
- The administrator can add information to the database without stopping the application.
- If implemented at Village, District and State level, the model will provide valuable information to other agencies and panchayats in particular.

Problems with the Use of Decision Support Systems

- User interfaces are often poorly designed
- Input data is not always objective or real

Data flows of a Decision Support System

- Incoming data
- Performed field operations and measurements and also the records of the farm
- Additional, usually contemporary information in the area of diseases and pests
- External guidance (human advisors)
- Outgoing data - plans and reports
- Data formats pose technical problems

A decision support system is a way to model data and make quality decisions based upon it. Making the right decision in business is usually based on the quality of the data and ability to sift through and analyze the data to find trends in which we can create solutions and strategies for. DSS or decision support systems are usually computer applications along with a human component that can sift through large amounts of data and pick between the many choices.

While many people think of decision support systems as a specialized part of a business, most companies have actually integrated this system into their day to day operating activities. For instance, many companies constantly download and analyze sales data, budget sheets and forecasts and they update their strategy once they analyze and evaluate the current results. Decision support systems have a definite structure in businesses, but in reality, the data and decisions that are based on it are fluid and constantly changing.

The key to decision support systems is to collect data, analyze and shape the data that is collected and then try to make sound decisions or construct strategies from analysis. Whether computers, databases or people are involved usually doesn't matter, however it is this process of taking raw or unstructured data, containing and collecting it and then using it to help aid decision making.

It is important to note that although computers and artificial intelligence is at work or in play with data, it is ultimately up to humans to execute these strategies or comprehend the data into a usable hypothesis.

Decision Support System does not have a universally accepted model, meaning that there are many theories waiting for supremacy in this broad field. Because of there are many working theories in the topic of DSS, there are many ways to classify DSS.

For instance, one of the DSS models available is with the relationship of the user in mind. This model takes into consideration passive, active and cooperative DSS models.

Decision support systems that just collect data and organize it effectively are usually called passive models, they do not suggest a specific decision and they only reveal the data. An active decision support system actually processes data and explicitly shows solutions based upon that data. While there are many systems that are able to be active, many organizations would be hard pressed to put all their faith into a computer model without any human intervention.

A cooperative decision support system is when data is collected, analyzed and then is provided to a human component which then can help the system revise or refine it. It means that both a human component and computer component work together to come up with the best solution.

Exercise to Practice

1. Define DSS
2. What are the components of DSS?

3. Types of DSS.
4. Scope of DSS.
5. Advantage of DSS.
6. Explain in detail about the use of DSS for farm management.

CHAPTER 10

Genetic Algorithms

Introduction

Genetic Algorithm simulates Darwinian evolutionary processes and naturally occurring genetic operations on chromosomes. In nature, chromosomes are character strings in nature's Base-4 alphabet. The four nucleotide bases are DNA molecule (strings) like Adenine (A), Guanine (G), Cytosine (C) and Thymine (T). The sequence of nucleotide bases is the chromosome string that determines the genome of a biological individual.

Eg: ATTCGTGC

A genetic algorithm (GA) is a search technique used in computing to find exact or approximate solutions to optimization and search problems. Genetic algorithms are categorized as global search heuristics. Genetic algorithms are a particular class of evolutionary algorithms (EA) that use techniques inspired by evolutionary biology such as inheritance, mutation, selection, and crossover.

Representation	Binary strings
Recombination	N-point or uniform
Mutation	Bitwise flipping with fixed probability
Parent selection	Fitness-Proportionate
Survivor selection	All children replace parents
Speciality	Emphasis on crossover

General Representation of Genetic Algorithm

Features of Genetic Algorithm

GA's are:

- Problem solvers
- Challenging technical puzzle
- Basis for machine learning
- Computational model for creativity and innovation
- Guiding philosophy
- Simulators of Darwinian theory

The four major steps in Genetic algorithm for solving a problem are:

1. Determine the representation schema
2. Determine the fitness measure
3. Determine the parameters and variables for controlling the algorithm
4. Determine the way of designating the result and the criterion for terminating a run.

Methodology

Genetic algorithms are implemented in a computer simulation in which a population of abstract representations (called chromosomes or the genotype of the genome) of candidate solutions (called individuals, creatures, or phenotypes) to an optimization problem evolves toward better solutions. Traditionally, solutions are represented in binary as strings of 0s and 1s, but other encodings are also possible. The evolution usually starts from a population of randomly generated individuals and happens in generations.

In each generation, the fitness of every individual in the population is evaluated, multiple individuals are stochastically selected from the current population (based on their fitness), and modified (recombined and possibly randomly mutated) to form a new population. The new population is then used in the next iteration of the algorithm. Commonly, the algorithm terminates when either a maximum number of generations has been produced, or a satisfactory fitness level has been reached for the population. If the algorithm has terminated due to a maximum number of generations, a satisfactory solution may or may not have been reached.

Steps in Executing the Genetic Algorithm

1. Randomly create an initial population of individual fixed length character strings.
2. Iteratively perform the following sub steps on the population of strings until the termination criteria met
   a) Evaluate fitness of each individual in the population.
   b) Create a new population of strings (selection method, apply the operators)
3. The best individual sting that appeared in any generation is designated as the result of genetic algorithm for the run. This result may represent a solution (or an approximate solution) to the problem.

The flowchart is given as:

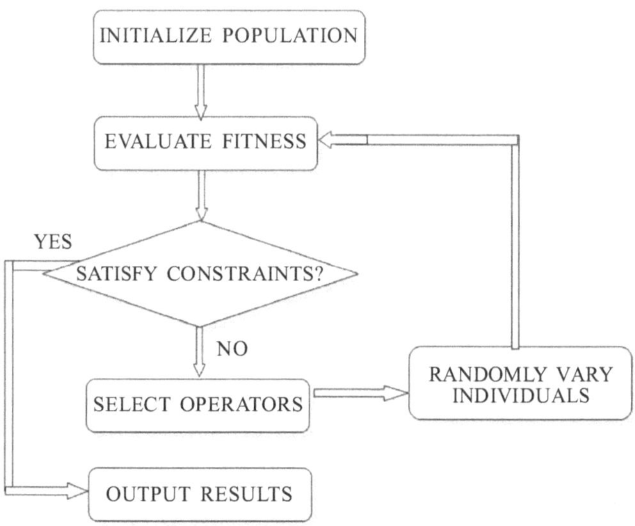

Genetic algorithms finds its application in bioinformatics, phylogenetics, computational science, engineering, economics, chemistry, manufacturing, mathematics, physics and other fields.

A typical genetic algorithm requires:

1. a genetic representation of the solution domain,
2. a fitness function to evaluate the solution domain.

A standard representation of the solution is as an array of bits. Arrays of other types and structures can be used in essentially the same way. The main property that makes these genetic representations convenient is that their parts are easily aligned due to their fixed size, which facilitates simple crossover operations. Variable length representations may also be used, but crossover implementation is more complex in this case. Tree-like representations are explored in genetic programming and graph-form representations are explored in evolutionary programming.

The fitness function is defined over the genetic representation and measures the quality of the represented solution. The fitness function is always problem dependent. For instance, in the knapsack problem one wants to maximize the total value of objects that can be put in a knapsack of some fixed capacity. A representation of a solution might be an array of bits, where each bit represents a different object, and the value of the bit (0 or 1) represents whether or not the object is in the knapsack. Not every such representation is valid, as the size of objects may exceed the capacity of the knapsack. The fitness of the solution is the sum of values of all objects in the knapsack if the representation is valid or 0 otherwise. In some problems, it is hard or even impossible to define the fitness expression; in these cases, interactive genetic algorithms are used.

Once we have the genetic representation and the fitness function defined, GA proceeds to initialize a population of solutions randomly, and then improve it through repetitive application of mutation, crossover, inversion and selection operators.

Initialization

Initially many individual solutions are randomly generated to form an initial population. The population size depends on the nature of the problem, but typically contains several hundreds or thousands of possible solutions. Traditionally, the population is generated randomly, covering the entire range of possible solutions (the search space). Occasionally, the solutions may be "seeded" in areas where optimal solutions are likely to be found.

Selection

During each successive generation, a proportion of the existing population is selected to breed a new generation. Individual solutions are selected through a fitness-based process, where fitter solutions (as measured by a fitness function) are typically more likely to be selected. Certain selection methods rate the fitness of each solution and preferentially select the best solutions. Other methods rate only a random sample of the population, as this process may be very time-consuming.

Most functions are stochastic and designed so that a small proportion of less fit solutions are selected. This helps keep the diversity of the population large, preventing premature convergence on poor solutions. Popular and well-studied selection methods include roulette wheel selection and tournament selection.

Reproduction (Cross – over and Mutation)

The next step is to generate a second generation population of solutions from those selected through genetic operators: crossover (also called recombination), and/or mutation.

For each new solution to be produced, a pair of "parent" solutions is selected for breeding from the pool selected previously. By producing a "child" solution using the above methods of crossover and mutation, a new solution is created which typically shares many of the characteristics of its "parents". New parents are selected for each new child, and the process continues until a new population of solutions of appropriate size is generated. Although reproduction methods that are based on the use of two parents are more "biology inspired", some research suggests more than two "parents" are better to be used to reproduce a good quality chromosome.

These processes results in the next generation population of chromosomes that is different from the initial generation. Generally the average fitness will have increased by this procedure for the population, since only the best organisms from the first generation are selected for breeding, along with a small proportion of less fit solutions.

Termination

This generational process is repeated until a termination condition has been reached. Common terminating conditions are:

- A solution is found that satisfies minimum criteria
- Fixed number of generations reached
- Allocated budget (computation time/money) reached
- The highest ranking solution's fitness is reaching or has reached a plateau such that successive iterations no longer produce better results
- Manual inspection
- Combinations of the above

Simple generational genetic algorithm pseudo codes are:

1. Choose the initial population of individuals
2. Evaluate the fitness of each individual in that population
3. Repeat on this generation until termination: (time limit, sufficient fitness achieved, etc.)
   a. Select the best-fit individuals for reproduction
   b. Breed new individuals through crossover and mutation operations to give birth to offspring
   c. Evaluate the individual fitness of new individuals
   d. Replace least-fit population with new individuals

Applications of Genetic Algorithm

GA's are useful in solving the real time problems like:

- Computer games (snake game)
- File allocation for a distributed system
- Travelling salesman problem (TSP):
  - routing of school buses, postal carriers and delivery trucks
  - routing of computer networks' parallel processing
- Artificial ant problem
- Knapsack problem

Artificial Ant Problem

- Problem deals with finding a minimal ant trial covering the maximum number of food particles placed in a 5x5 grid - "food foraging" technique.
- Used in traffic management systems where shortest path is to be found to avoid traffic signals.

Encoding

- With a population of possible solutions, we encode a path as an array of elements where each element can be 1,2,3 or 4 etc.
- Boundaries of the grid are not being crossed and the ant does not retrace its path.

Fitness Measure

- Fitness of each possible solution is based on
- the number of grids traversed (and)
- the number of food particles eaten
- Gap penalty is also introduced.
- Fitness = no. of food particles eaten- gap penalties.
- This type is the raw fitness measure.

Cross-over operation

- Starts with two parents from initial population and produces two new offsprings.
- Single point crossover mechanism is employed.
- Check for consistency.
- If any inconsistency occurs, the child is mutated again.

Mutation

- It refers to either reintroduce or eliminate a gene by introducing changes in the structure in the population.
- Once again, the off springs are checked for consistencies before they go to the next stage.

Selection

- Here, tournament selection is used.
- It involves choosing the best chromosomes among the parent and child populations.

For example, consider a 5 x 5 grid where the ant has to cover all the food particles without retracing its path. This problem can be solved using Genetic Algorithm and the illustration can be given as follows:

Example - 5 X 5 Grid

Initial Population

1	6	7	12	13	14

Fitness = 3 - 3 = 0

1	2	7	12	13	14

Fitness = 3 - 4 = -1

1	6	7	8	13	14

Fitness = 3 - 3 = 0

1	7	12	13	18	23

Fitness = 3 - 4 = -1

1	2	3	12	13	14

Fitness = 2 - 4 = 2

5	10	15	14	19	25

Fitness = 2 - 4 = -2

2	7	8	9	14	19

Fitness = 2 - 4 = 2

6	7	8	13	14	19

Fitness = 3 - 3 = 0

6	7	12	13	18	19

Fitness = 3 - 3 = 0

2	7	12	13	18	23

Fitness = 2 - 4 = -2

Cross over operation

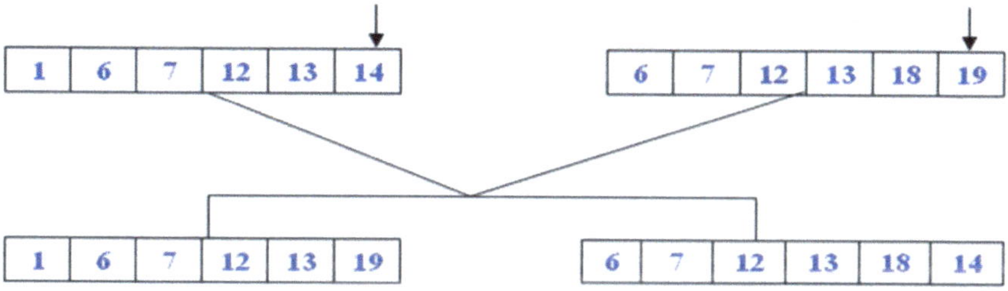

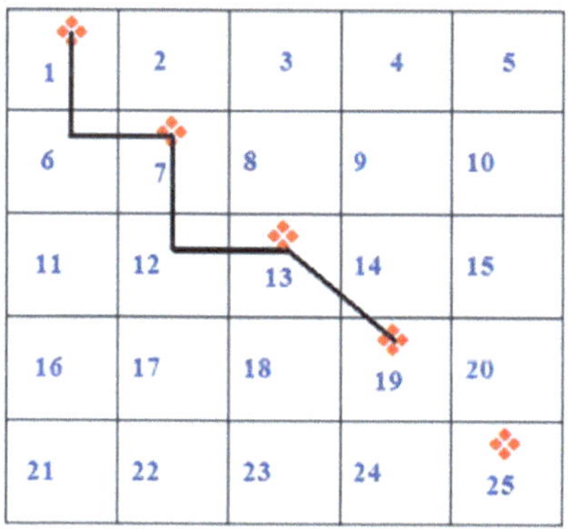

Fitness = 4 - 2 = 2

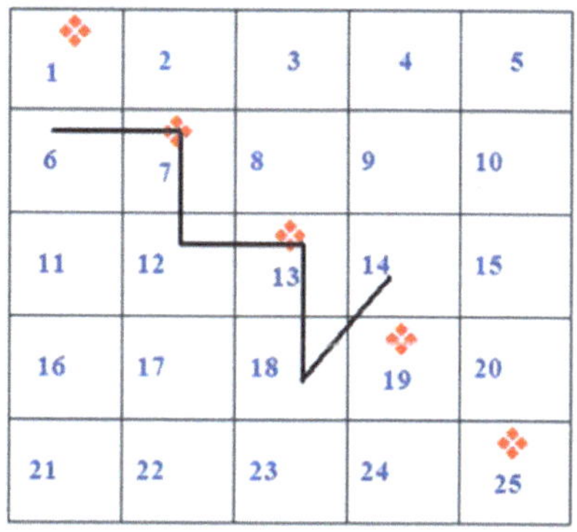

Fitness = 2 - 4 = -2

Mutation

Mutation - Generation I

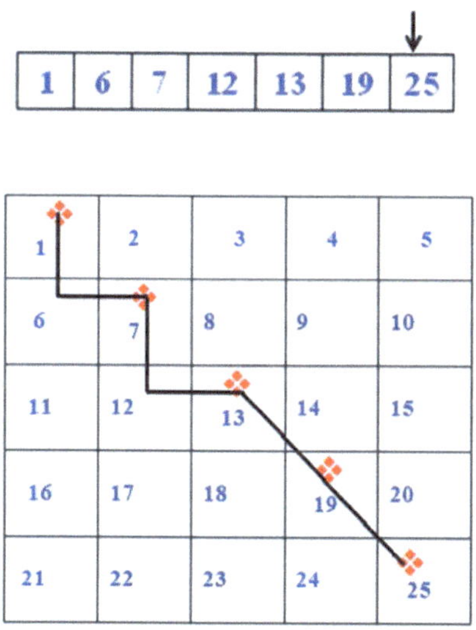

Fitness = 5 - 2 = 3

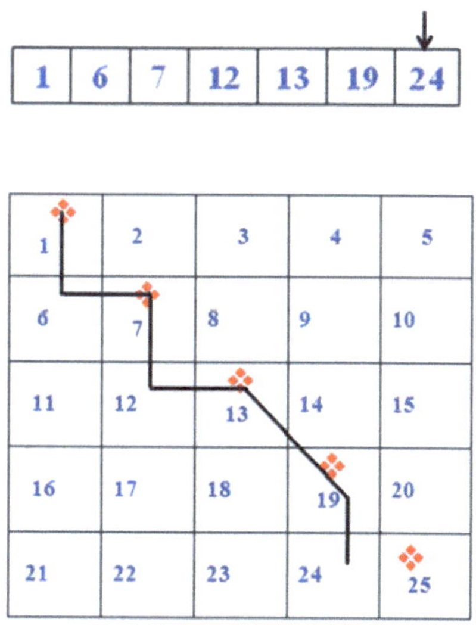

Fitness = 4 - 3 = 1

Mutation - Contd...

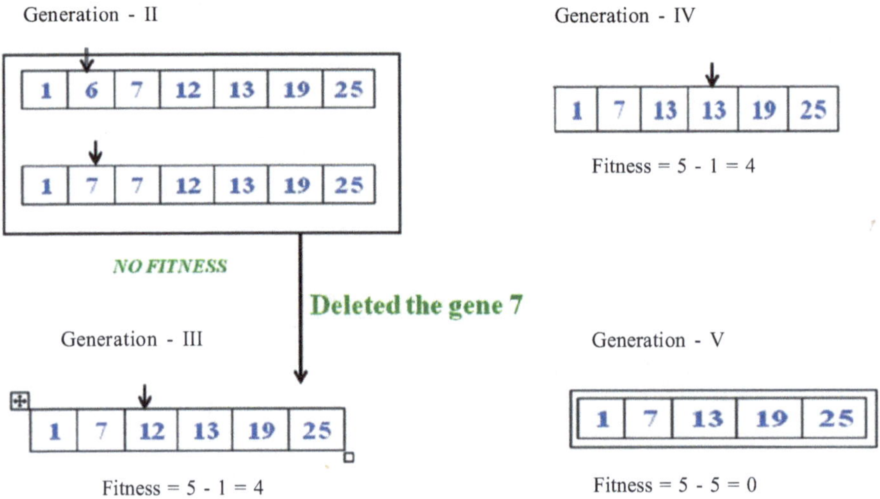

Generation V - Final Solution

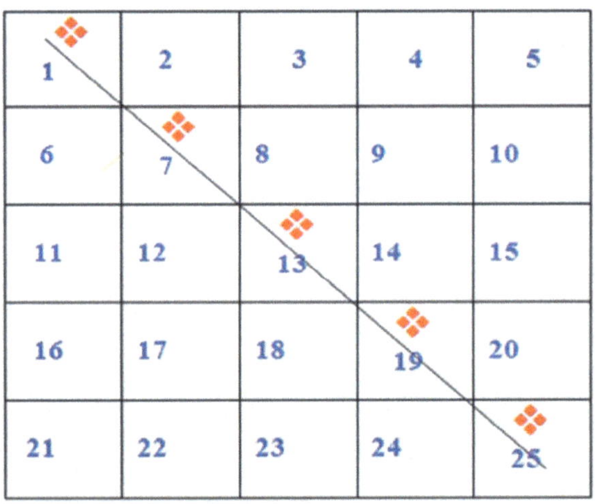

Fitness = 5 - 0 = 5
Final solution

Observations

There are several general observations about the generation of solutions via a genetic algorithm:

- Repeated fitness function evaluation for complex problems are often the most prohibitive and limiting segment of artificial evolutionary algorithms. Finding optimal solution to complex high dimensional, multimodal problems often requires very expensive fitness function evaluations. In real world problems such as structural optimization problems, one single function evaluation may require several hours to several days of complete simulation. Typical optimization method can not deal with such a type of problem. In this case, it may be necessary to go for an exact evaluation and use an approximated fitness that is computationally efficient. It is apparent that the combination of approximate models may be one of the most promising approaches to use Genetic Algorithms to solve complex real life problems.

- The "better" is only in comparison to other solutions. As a result, the stop criterion is not clear.
- In many problems, GAs may have a tendency to converge towards local optima or even arbitrary points rather than the global optimum of the problem. This means that it does not "know how" to sacrifice short-term fitness to gain longer-term fitness. The likelihood of this occurring depends on the shape of the fitness landscape: certain problems may provide an easy ascent towards a global optimum; others may make it easier for the function to find the local optima. This problem may be alleviated by using a different fitness function, increasing the rate of mutation, or by using selection techniques that maintain a diverse population of solutions, although the No Free Lunch theorem proves that there is no general solution to this problem. A common technique to maintain diversity is to impose a "penalty", wherein, any group of individuals of sufficient similarity (radius) have a penalty added, which will reduce the representation of that group in subsequent generations, permitting other (less similar) individuals to be maintained in the population. This however, may not be effective, depending on the landscape of the problem. Diversity is important in genetic algorithms (and genetic programming) because crossing over a homogeneous population does not yield new solutions. In evolution strategies and evolutionary programming, diversity is not essential because of a greater reliance on mutation.
- Operating on dynamic data sets is difficult, as genomes begin to converge early on towards solutions which may no longer be

valid for later data. Several methods have been proposed to remedy this by increasing genetic diversity somehow and preventing early convergence, either by increasing the probability of mutation when the solution quality drops, or by occasionally introducing entirely new, randomly generated elements into the gene pool.

Again, evolution strategies and evolutionary programming can be implemented with a so-called "comma strategy" in which parents are not maintained and new parents are selected only from offspring. This can be more effective on dynamic problems.

- GAs cannot effectively solve problems in which the only fitness measure is a single right/wrong measure (like decision problems), as there is no way to converge on the solution (no hill to climb). In these cases, a random search may find a solution as quickly as a GA. However, if the situation allows the success/failure trial to be repeated giving (possibly) different results, then the ratio of successes to failures provides a suitable fitness measure.
- Selection is clearly an important genetic operator, but opinion is divided over the importance of crossover versus mutation. Some argue that crossover is the most important, while mutation is only necessary to ensure that potential solutions are not lost. Others argue that crossover in a largely uniform population only serves to propagate innovations originally found by mutation, and in a non-uniform population crossover is nearly always equivalent to a very large mutation (which is likely to be catastrophic). Often, GAs can rapidly locate good solutions, even for difficult search spaces. The same is of course also true for evolution strategies and evolutionary programming.
- For specific optimization problems and problem instances, other optimization algorithms may find better solutions than genetic algorithms (given the same amount of computation time). Alternative and complementary algorithms include evolution strategies, evolutionary programming, simulated annealing, Gaussian adaptation, hill climbing, and swarm intelligence (e.g.: ant colony optimization, particle swarm optimization) and methods based on integer linear programming.
- As with all current machine learning problems it is worth tuning the parameters such as mutation probability, recombination

probability and population size to find reasonable settings for the problem class being worked on.

A very small mutation rate may lead to genetic drift . A recombination rate that is too high may lead to premature convergence of the genetic algorithm. A mutation rate that is too high may lead to loss of good solutions unless there is superior selection. There are theoretical but not yet practical upper and lower bounds for these parameters that can help guide selection.

- The implementation and evaluation of the fitness function is an important factor in the speed and efficiency of the algorithm.

Exercise to Practice

1. What is genetic algorithm?
2. How a Genetic Algorithm Works?
3. Why use a genetic algorithm?
4. Define the following:

   Selection

   Mutation

   Cross-over operation
5. What are the Steps in executing the Genetic Algorithm?
6. Give some applications of Genetic Algorithm in Agriculture.

CHAPTER 11

Programming Language - Visual Basic

A Brief Description of Visual Basic

Visual Basic is a high level programming language which was evolved from the earlier DOS version called Basic. Basic means Beginners' All-purpose Symbolic Instruction Code. It is a very easy programming language to learn. Visual Basic is a Visual and events driven Programming Language. These are the main divergence from the old Basic. In Basic, programming is done in a text-only environment and the program is executed sequentially. In VB, programming is done in a graphical environment. In the old Basic, we write program codes for each graphical object we wish to display it on screen, including its position and its color. However, In Visual Basic, we just need to drag and drop any graphical object anywhere on the form, and we can change its color any time using the properties windows.

On the other hand, users may click on certain object randomly, so each object has to be programmed independently to be able to response to those actions (events). Therefore, a VB Program is made up of many subprograms, each has its own program codes, and each can be executed independently and at the same time each can be linked together in one way or another.

Types of Application in VB

1. Standard Exe
2. Active Exe, ActiveX DLL
3. ActiveX control
4. ActiveX Document Exe
5. VB Application wizard
6. ADDIN
7. VB wizard manager
8. VB Enterprise Edition controls

Form

A form is a window or a dialog box overlaid on the screen and used as a display area or for getting information and instructions from the user.

Designing a Form

Two windows

- Form itself
- A code window

Properties

Properties are a collection of items that define a particular form or a controls its appearance and its operations.

Eg: Appearance, Alignment, Backcolor etc.

Data Access

Data Access refers to the way a programmer interacts with the back end or the DBMS where most of the data are stored.

DAO (Data Access Objects)

RDO (Remote Data Objects)

ADO (Active Data objects)

Objects

Objects are entities which can be manipulated like forms, controls etc., An object is a piece of software that has properties and functions that can be manipulated. A window is an object. It has properties: size, color, position on the screen, etc. The window has functions, also called methods, that can be manipulated: change the size, move it around, open it and close it.

Controls

Controls are objects that are put on the forms to display information or get a response from the user.

Projects

Project is a collection of all forms modules and controls that make up the application.

Creating the Project

To create a visual basic project click Start – >Microsoft Visual Basic 6.0 screen will appear as shown below:

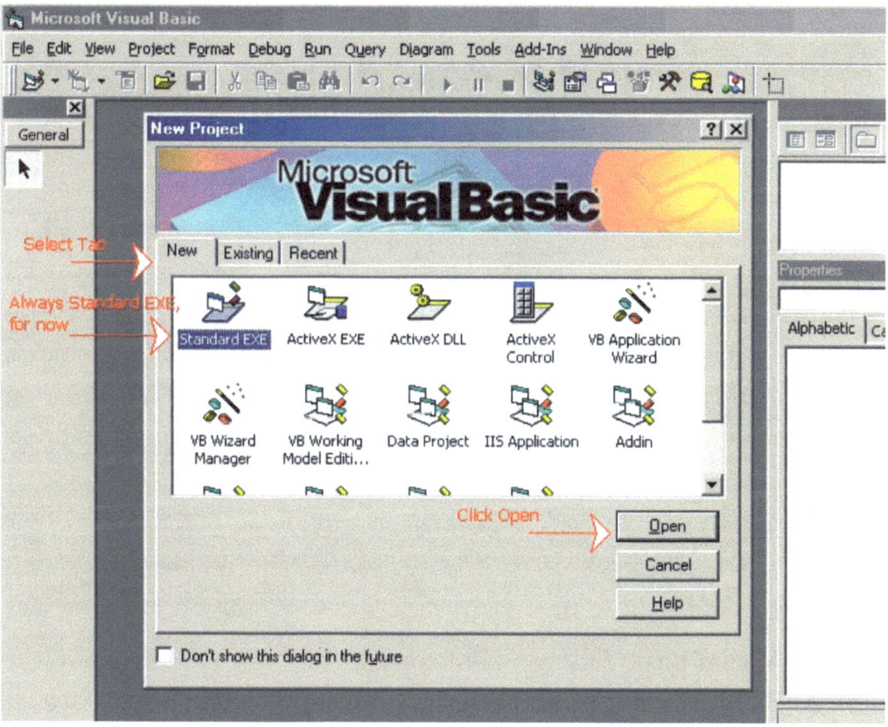

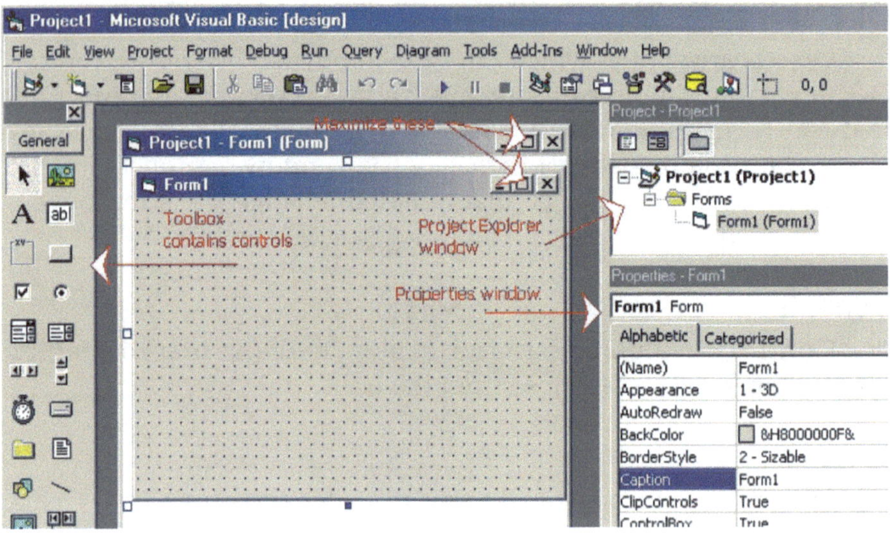

For example we need 6 **labels** and 2 **command buttons**. Each one of these objects on a Form is called a **control**. To get a control in the Form, click **Toolbox.** Controls are placed in the Form as shown below:

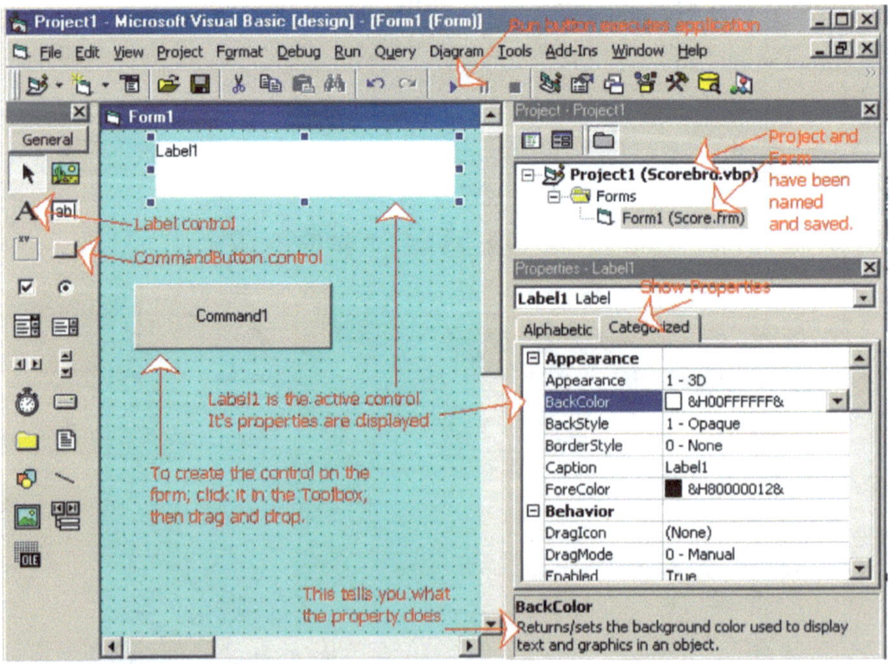

The Visual Basic Environment consists of

- A Blank Form for you to design your application's interface.

- The Project window which displays the files that are created in your application.
- The Properties window which displays the properties of various controls and objects that are created in your applications.

It also includes a Toolbox that consists of all the controls essential for developing a VB Application. Controls are tools such as text box, command button, label, combo box, picture box, image box, timer and other objects that can be dragged and drawn on a form to perform certain tasks according to the events associated with them.

Choose from the Visual Basic menu File->New Project as shown below:

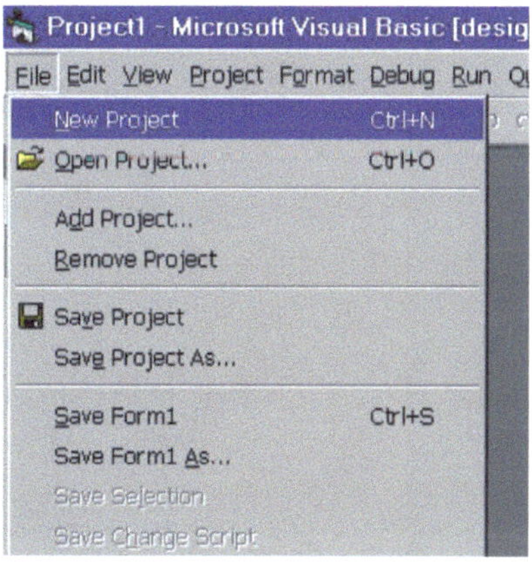

Dialog box opens as shown below

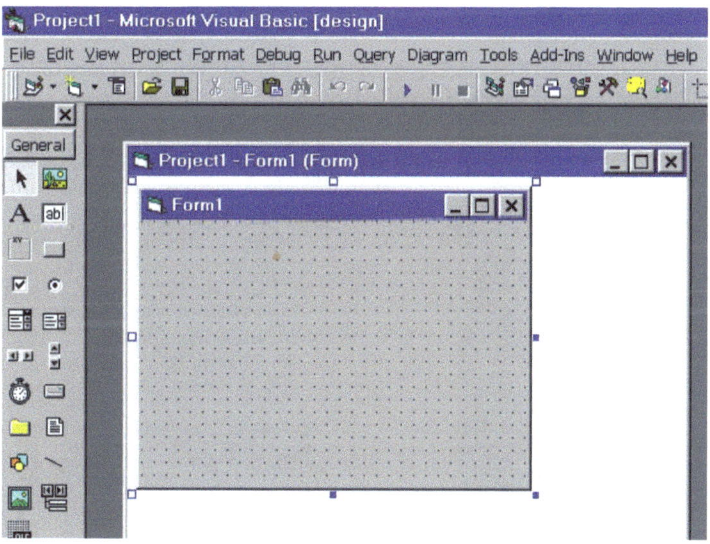

Click on the Play button in the toolbar

Output window appears as shown below

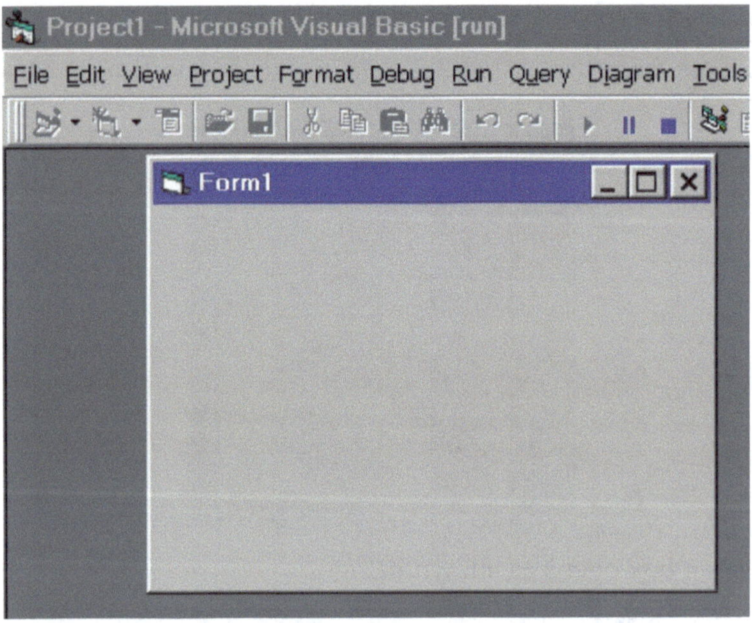

To stop the program from running, we can click on the Stop button in the Toolbar, or click the form's X button.

Steps in Building a Visual Basic Application

Generally, there are three basic steps in building a VB application. The steps are as follows:

Step 1: Design the interface.

Step 2: Set Properties of the controls (Objects)

Step 3: Write the events' procedures

Building a Visual Basic Application

In Microsoft Visual Basic, a default form Form1 will be available in new project. Now, double click on form1, the source code window for form1 as shown in figure will appear. The top of the source code window consists of a list of objects and their associated events or procedures. The object displayed is Form and the associated procedure is Load.

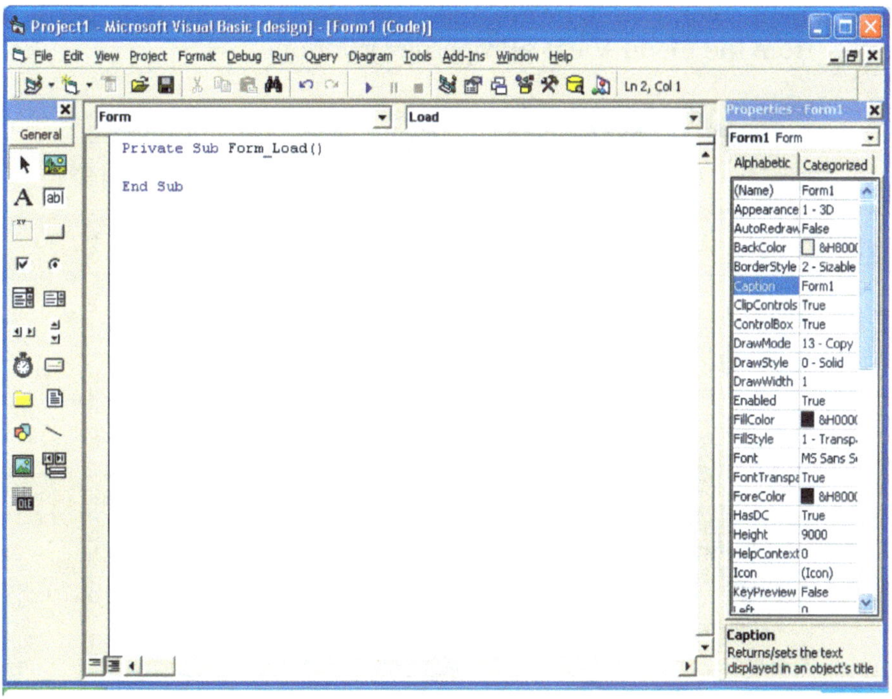

Example

```
Private Sub Form_Load ( )
    Form1.show
    Print "Welcome to Visual Basic l"
End Sub
```

Writing the Codes

Each control or object in Visual Basic can usually run many kinds of events or procedures; these events are listed in the dropdown list in the code window that is displayed when you double-click on an object and click on the procedures' box. For each event, you need to write an event procedure so that an action or a series of actions can be performed.

To start writing an event procedure, you need to double-click an object. For example, if you want to write an event procedure when a user clicks a command button, you double-click on the command button and an event procedure will appear.

Properties in Visual Basic

Every component (form is component for example) has properties, that determine its look and its functioning. Properties can be the component color, width, height and more. To see the form properties, select from the menu View->Properties Window

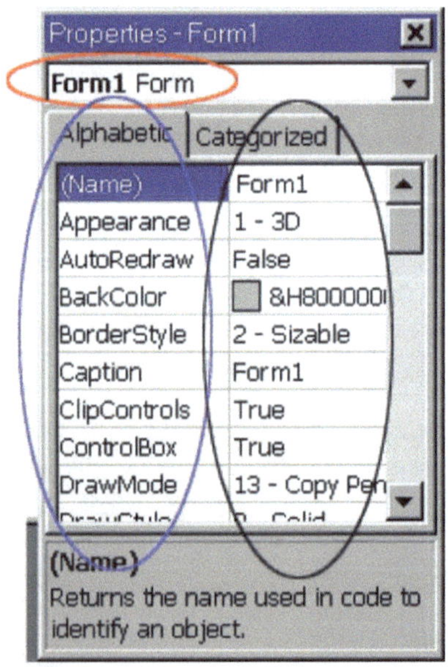

The properties window looks like this:

In the above diagram the Form name is Form1. The form property names are appearance property AutoRedraw property and more.

The form's properties values are form's Appearance property is 1 - 3D, and so on. We can also change the form's properties. To change the Caption property, simply click on the Caption property in the form's properties names column.

As we see, the current Caption is Form1. Delete the "Form1" text and type instead of it "Hello".

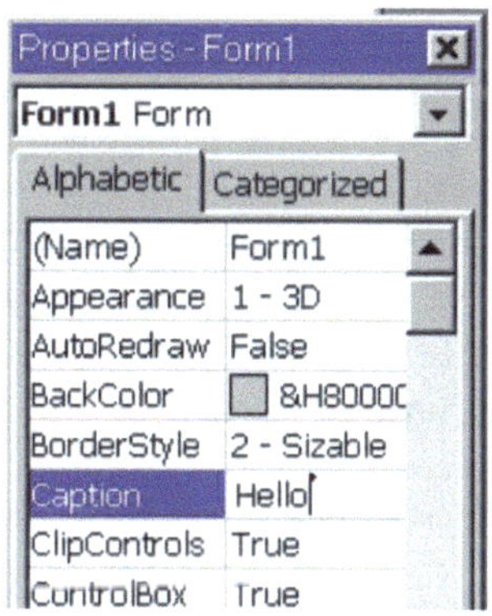

Now run the program using the Play button. You will see that the text on the form's Title bar is Hello.

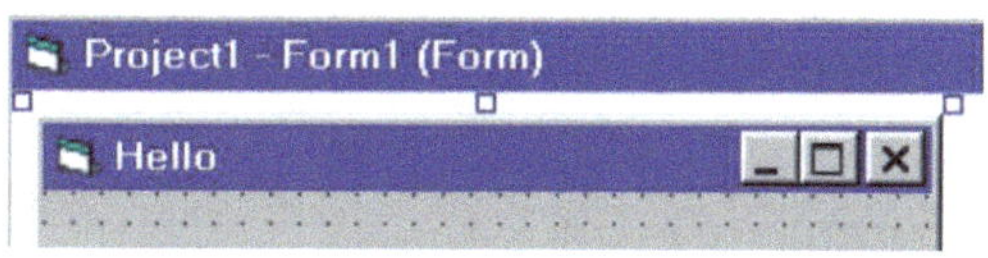

Adding Controls to the Form

There are many controls such as Buttons, Text Boxes, Pictures, Scroll Bars and more. To see the Toolbox, Select from the menu View->Toolbox, we can view Toolbox

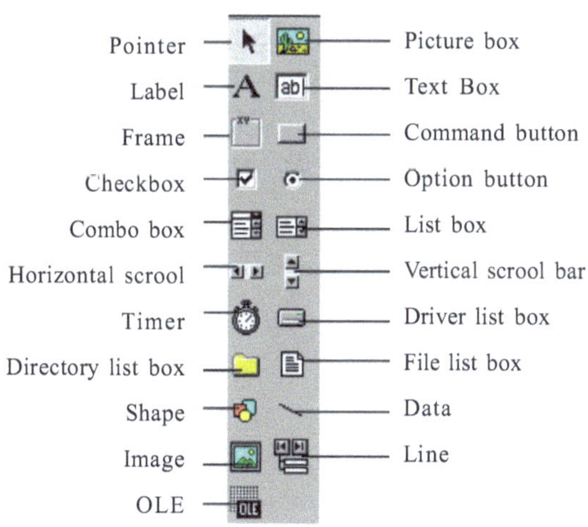

Text Box

The text box is the standard control for accepting input from the user as well as to display the output. It can handle string (text) and numeric data but not images or pictures. String in a text box can be converted to a numeric data by using the function Val (text).

Label

The label is a very useful control for Visual Basic, as it is not only used to provide instructions and guides to the users, it can also be used to display outputs. One of its most important properties is **Caption**. Using the syntax **label.Caption**, it can display text and numeric data. We can change its caption in the properties window and also at runtime.

Command Button

The command button is one of the most important controls as it is used to execute commands. It displays an illusion that the button is pressed when the user click on it. The most common event associated with the command button is the Click event, and the syntax for the procedure is

```
Private Sub Command1_Click ()
Statements
End Sub
```

Picture Box

The Picture Box is one of the controls that is used to handle graphics. We can load a picture at design phase by clicking on the picture item in the properties window and select the picture from the selected folder. We can load the picture at runtime using the **LoadPicture** method or with the statement given below.

Picture1.Picture=LoadPicture ("C:\VB program\ Images\ orange.gif")

Image Box

Image Box is another control that handles images and pictures. It functions almost identically to the picture box. However, there is one major difference, the image in an Image Box is stretchable, which means it can be resized. This feature is not available in the Picture Box. Similar to the Picture Box, it can also use the LoadPicture method to load the picture.

List Box

The function of the List Box is to present a list of items where the user can click and select the items from the list. In order to add items to the list, we can use the AddItem method. For example, if you wish to add a number of items to list box1, we can key in the following statements

```
Private Sub Form_Load ( )
List1.AddItem "rice"
List1.AddItem "wheat"
List1.AddItem "orange"
List1.AddItem "apple"
End Sub
```

The items in the list box can be identified by the ListIndex property, the value of the ListIndex for the first item is 0, the second item has a List Index1, and the second item has a List Index2 and so on.

Combo Box

The function of the Combo Box is also to present a list of items where the user can click and select the items from the list. However, the user needs to click on the small arrowhead on the right of the combo box to see the items which are presented in a drop-down list. In order to add items to the list, we can also use the AddItem method. For example, if we wish to add a number of items to Combo box1, we can key in the following statements.

```
Private Sub Form_Load ( )
Combo1.AddItem "Item1"
Combo1.AddItem "Item2"
Combo1.AddItem "Item3"
Combo1.AddItem "Item4"
End Sub
```

Check Box

Check Box control lets the user selects or unselects an option. When the Check Box is checked, its value is set to 1 and when it is unchecked, the value is set to 0. We can include the statements Check1.Value=1 to mark the Check Box and Check1.Value=0 to unmark the Check Box, as well as use them to initiate certain actions. For example, the program will change the background color of the form to red when the check box is unchecked and it will change to blue when the check box is checked.

```
Private Sub Command1_Click()
If Check1.Value = 1 And Check2.Value = 0 Then
MsgBox "Apple is selected"
ElseIf Check2.Value = 1 And Check1.Value = 0 Then
MsgBox "Orange is selected"
Else
MsgBox "All are selected"
End If
End Sub
```

Option Box

The Option Box control also lets the user selects one of the choices. However, two or more Option Boxes must work together because as one of the Option Boxes is selected, the other Option Boxes will be unselected. In fact, only one Option Box can be selected at one time. When an option box is selected, its value is set to "True" and when it is unselected, its value is set to "False". In the following example, the shape control is placed in the form together with six Option Boxes. When the user clicks on different option boxes, different shapes will appear. The values of the shape control are 0, 1, and 2,3,4,5 which will make it appear as a rectangle, a square, an oval shape, a rounded rectangle and a rounded square respectively.

```
Private Sub Option1_Click ( )
Shape1.Shape = 0
End Sub

Private Sub Option2_Click()
Shape1.Shape = 1
End Sub

Private Sub Option3_Click()
Shape1.Shape = 2
End Sub
```

```
Private Sub Option4_Click()
Shape1.Shape = 3
End Sub

Private Sub Option5_Click()
Shape1.Shape = 4
End Sub

Private Sub Option6_Click()
Shape1.Shape = 5
End Sub
```

Drive List Box

Drive ListBox is for displaying a list of drives available in your computer. When you place this control into the form and run the program, we can select different drives from your computer.

Directory List Box

The Directory List Box is for displaying the list of directories or folders in a selected drive. When we place this control into the form and run the program, we can select different directories from a selected drive in your computer.

File List Box

The File List Box is for displaying the list of files in a selected directory or folder.

Events

An action recognized by an object, such as clicking the mouse or pressing a key and for which we can write, code to respond is known as an event eg: Click, Change, and Activate. Visual Basic is "Event Driven" language. Everything that happens, launch an event. When we move the mouse, "MouseMove" event has been launched. When we pressed a key? The "KeyPress" event has been launched.

Visual Basic Data Types

Visual Basic classifies the information into two major data types, they are the numeric data types and the non-numeric data types.

Numeric Data types

Numeric data types are types of data that consist of numbers, which can be computed mathematically with various standard operators such as add, minus, multiply, divide and more. Examples of numeric data types are examination marks, height, weight, the number of students in a class, height of plant, yield and others.

In Visual Basic, numeric data are divided into 7 types, depending on the range of values they can store. Calculations that only involve round figures or data that does not need precision can use Integer or Long integer in the computation. Programs that require high precision calculation need to use Single and Double decision data types, they are also called floating point numbers. For currency calculation, we can use the currency data types.

Type	Storage	Range of Values
Byte	1 byte	0 to 255
Integer	2 bytes	-32,768 to 32,767
Long	4 bytes	-2,147,483,648 to 2,147,483,648
Single	4 bytes	-3.402823E+38 to -1.401298E-45 for negative values 1.401298E-45 to 3.402823E+38 for positive values.
Double	8 bytes	-1.79769313486232e+308 to -4.94065645841247E-324 for negative values 4.94065645841247E-324 to 1.79769313486232e+308 for positive values.
Currency	8 bytes	-922,337,203,685,477.5808 to 922,337, 203,685, 477.5807
Decimal	12 bytes	+/- 79,228,162,514,264,337,593,543,950,335 if no decimal is use +/- 7.9228162514264337593543950335 (28 decimal places).

Numeric Data Types

Non-numeric Data Types

Nonnumeric data types are data that cannot be manipulated mathematically using standard arithmetic operators. The non-numeric data comprises text or string data types, the Date data types, the Boolean data types that store only two values (true or false), Object data type and Variant data type .

Data Type	Storage	Range
String (fixed length)	Length of string	1 to 65,400 characters
String (variable length)	Length + 10 bytes	0 to 2 billion characters
Date	8 bytes	January 1, 100 to December 31, 9999
Boolean	2 bytes	True or False
Object	4 bytes	Any embedded object
Variant (numeric)	16 bytes	Any value as large as Double
Variant (text)	Length+22 bytes	Same as variable-length string

Nonnumeric Data Types

Managing Variables

Variables are areas allocated by the computer memory to hold data. Each variable must be given a name. To name a variable in Visual Basic, we follow the set of rules.

The following are the rules when naming the variables in Visual Basic

- It must be less than 255 characters
- No spacing is allowed
- It must not begin with a number
- Period is not permitted

Examples of valid and invalid variable names are displayed in Table

Valid Name	Invalid Name
My_Dog	My.Dog
ThisMonth	1Newcow
Long_Age_Can_beUSE	He&HisMother *& is not acceptable

Declaring Variables

In Visual Basic, one needs to declare the variables before using them by assigning names and data types. They are normally declared in the general section of the code windows using the **Dim** statement.

The format is as follows:

Dim Variable Name As Data Type

Example

Dim password As String

Dim yourName As String

Dim firstname As Integer

Dim secondname As Integer

Dim totalmark As Integer

Dim toDate As Date

You may also combine them in one line, separating each variable with a comma, as follows:

Dim password As String, yourName As String, firstname As Integer,.............

If data type is not specified, Visual Basic will automatically declare the variable as a Variant. For string declaration, there are two possible formats, one for the variable-length string and another for the fixed-length string.

Dim VariableName as String * *n*, where n defines the number of characters the string can hold.

Example

Dim yourName as String * 10

yourName can holds no more than 10 Characters.

Constants

Constants are different from variables in the sense that their values do not change during the running of the program.

Declaring a Constant

The format to declare a constant is

Const Constant Name As Data Type = Value

Example 5.3

Const Pi As Single=3.142

Const Tem As Single=35

Const Score As Single=101

Assigning Values to Variables

After declaring various variables using the Dim statements, we can assign values to those variables. The general format of an assignment is

Variable=Expression

The variable can be a declared variable or a control property value. The expression could be a mathematical expression, a number, a string, a Boolean value (true or false) and more.

String Functions

List of the basic functions are

- *Len (string):* returns the length of string, the number of characters it contains.
- *Left (string, number):* returns the number of characters specified by number from the left end of string.
- *Right (string, number):* returns the number of characters specified by number from the right end of string.
- *Mid (string, position, number):* returns the number of characters specified by number starting at character number position from the left end of string.
- *InStr (string1, string2):* returns the position of string2 in string1 - returns 0 if string2 is not found in string1.
- *LTrim (string), RTrim (string) and Trim (string):* returns string with non-significant spaces removed from the left, the right or both respectively.
- *LCase (string), UCase (string):* returns string converted to lower-case or upper-case.

Operators in Visual Basic

To compute inputs from users and to generate results, we use various operators. In operator in Visual Basic are

- Arithmetic Operators
- Conditional Operators
- Logical Operators

Arithmetic Operators

Arithmetic operators are given below

Operator	Meaning
^	Exponential
*	Multiplication
/	Division
Mod	Modulus(return the remainder from an integer division)
\	Integer Division(discards the decimal places)
+ or &	String concatenation

Conditional Operators

Conditional operators are very powerful tools, they let the Visual Basic program to compare data values. The Conditional operators are

Operator	Meaning
=	Equal to
>	More than
<	Less Than
>=	More than and equal
<=	Less than and equal
<>	Not Equal to

Logical Operators

Logical operators are given below

Operator	Meaning
And	Both sides must be true
or	One side or other must be true
Xor	One side or other must be true but not both
Not	Negates truth

Controlling Program Flow

If.....Then.....Else Statements

To control the Visual Basic program flow, we use **If...Then...Else** statement together with the conditional operators and logical operators.

Syntax

The general format for the if...then...else statement is

If conditions Then

VB expressions

Else

VB expressions

End If

Select Case....End select statement

Select Case control structure is slightly different from the If....ElseIf control structure. The difference is that the Select Case control structure basically make decision on one expression or dimension (for example the examination grade) while the If ...ElseIf statement control structure may evaluate only one expression, each If....ElseIf statement may also compute entirely different dimensions. Select Case is preferred when there exists many different conditions.

Syntax

The format of the Select Case control structure are

```
Select Case expression
Case value1
   Block of one or more VB statements
Case value2
   Block of one or more VB Statements
Case value3
   ...
   ...
Case Else
   Block of one or more VB Statements
End Select
```

Example

```
Dim grade As String
Private Sub Compute_Click( )
```

```
grade = txtgrade.Text
Select Case grade
Case "A"
   result.Caption="High  Distinction"
Case "A-"
   result.Caption="Distinction"
Case "B"
   result.Caption="Credit"
Case "C"
   result.Caption="Pass"
Case Else
   result.Caption="Fail"
End Select
End Sub
```

Looping

Visual Basic allows a procedure to be repeated many times as long as the processor until a condition or a set of conditions is fulfilled. This is generally called looping. Looping is a very useful feature of Visual Basic because it makes repetitive works easier. There are two kinds of loops in Visual Basic, they are Do...Loop and the For.......Next loop

Do while ....Loop statement

Syntax

```
Do while condition
   Block of one or more VB statement
Loop
```

Example

```
Counter =1
Do while counter < 101
Counter = counter +1
Loop
```

Do ...Loop while statement

Syntax

Do

Block of one or more VB statement

Loop while condition

Example

Counter = 200

Do

Text1.text = str(counter)

Counter = counter + 1

Loop while counter < 50

For ... Next loop statement

Syntax

For counter = startNumber to endNumber step increment

Example

For I=1 to 50

Text1.text = str(I)

Msgbox "Number display"

Next

The InputBox( ) Function

An InputBox( ) function will display a message box where the user can enter a value or a message in the form of text. The format is

myMessage=InputBox(Prompt, Title, default_text, x-position, y-position)

myMessage is a variant data type but typically it is declared as string, which accept the message input by the users. The arguments are explained as follows:

- Prompt - The message displayed normally as a question asked.
- Title - The title of the Input Box.
- default-text - The default text that appears in the input field where users can use it as his intended input or he may change to the message he wish to key in.

- x-position and y-position - the position or the coordinate of the input box.

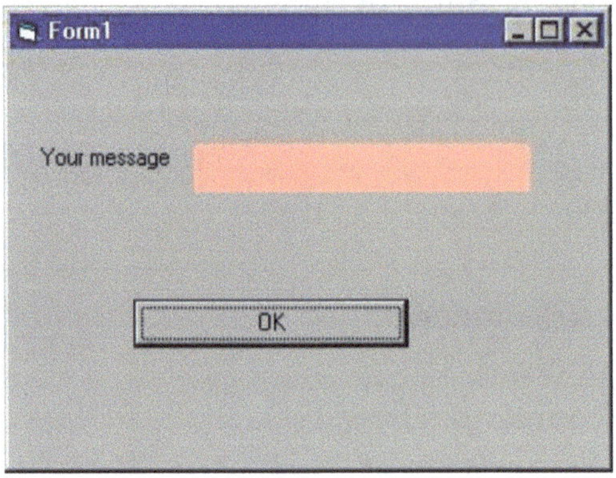

Example

The procedure for the OK button

```
Private Sub OK_Click()
Dim userMsg As String
userMsg = InputBox("What is your message?", "Message Entry
Form", "Enter your messge here", 500, 700)
If userMsg <> "" Then
message.Caption = userMsg
Else
message.Caption = "No Message"
End If
End Sub
```

When a user click the OK button, the input box as shown in Figure below will appear. After user entering the message and click OK, the message will be displayed on the caption, if he click Cancel, "No message" will be displayed.

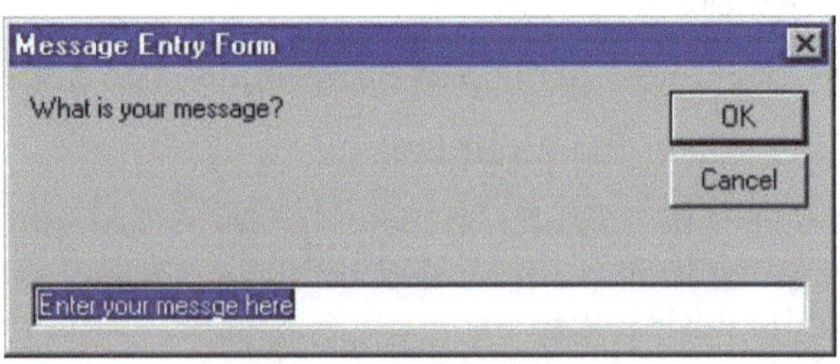

Methods

OOPS allows one to create data encapsulation ie., all the data and code associated with the data and code associated with the data members is grouped together to form an entity called class. All the code is written in various functions of the class. These functions are called methods of class.

Methods are

- Clear
- Add item, Remove item

Graphic methods are

1. Line - Draw a line, rectangle or filled in box
2. Circle - Draws a circle, ellipse or arc

Procedure

Simplify programming tasks by breaking programs into smaller logical components. These components are called procedure.

Sub updtform( )

Function Getcoord( )

Calling procedures

Call myproc (Argument1, Argumnet2)

(or)

myproc(Argument1, Argumnet2)

Procedures in forms

Call form1.somesub(arguments)

Control Array

A control array is a group of controls that share the same event procedures. The maximum index in control array is 32767.

Number(n)

0=Number(0)

1= Number(1)

2= Number(2)

3=Number(3)

Arrays

An array is **an indexed variable**, it is one variable with many parts, each part being referenced by an index number. The index number being numeric, it can be manipulated by loop statements, incremented, decremented, etc. An array can contain any valid data type and Variant type. An array is declared like other variable, with the addition of an index:

Dim Department(6) As String

will declare an array of 7 elements of the String type (we assume that it will be 7 Department names). The only problem with this declaration is that the index goes from 0 to 6. So, if we want the name of the sixth Department you have to specify Department(5).

Multidimensional Arrays

A multidimensional array is an array that looks like a table. It can also be refered to an array of arrays. Here is an example of a two-dimensional array.

Static iArray(1 To 2, 1 To 3)

All the elements list out are:

iArray(1,1), iArray(1,2), iArray(1,3), iArray(2,1), iArray(2,2), iArray(2,3)

In table it would look like this:

	1	2	3
1	iArray (1,1)	iArray (1,2)	iArray (1,3)
2	iArray (2,1)	iArray (2,2)	iArray (2,3)

Most of the time we use one dimensional arrays but some of the time it will be helpful to use 2 or even 3. More then 3 is not always a good idea because it tends to be hard to debug.

Sample Programs

i. The Interface

Coding

```
Dim studentName(10) As String
Dim num As Integer
Private Sub addName()
```

```
For num = 1 To 10
studentName(num) = InputBox("Enter the student name", "Enter Name", "", 1500, 4500)
If studentName(num) <> "" Then
Form1.Print studentName(num)
Else
End
End If
Next
End Sub
Private Sub Exit_Click()
End
End Sub
Private Sub Start_Click()
Form1.Cls
addName
End Sub
```

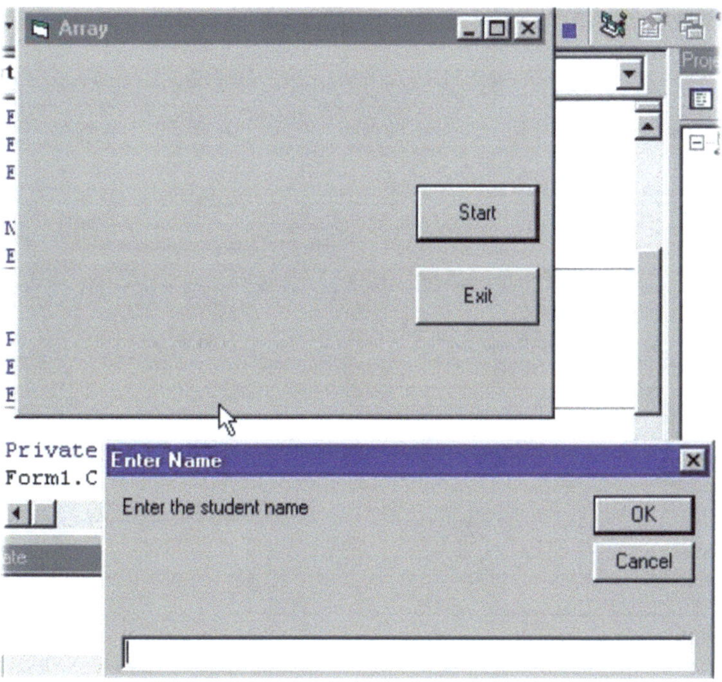

The above program accepts data entry through an input box and displays the entries in the form itself. This program will allows a user to enter 10 names each time. Click on the start button.

ii. The Interface

Coding

```
Dim studentName(10) As String
Dim num As Integer
Private Sub addName( )
For num = 1 To 10
studentName(num) = InputBox("Enter the student name")
List1.AddItem studentName(num)
Next
End Sub
Private Sub Start_Click()
addName
End Sub
```

The above program accepts data entries through an InputBox and displays the items in a list box.

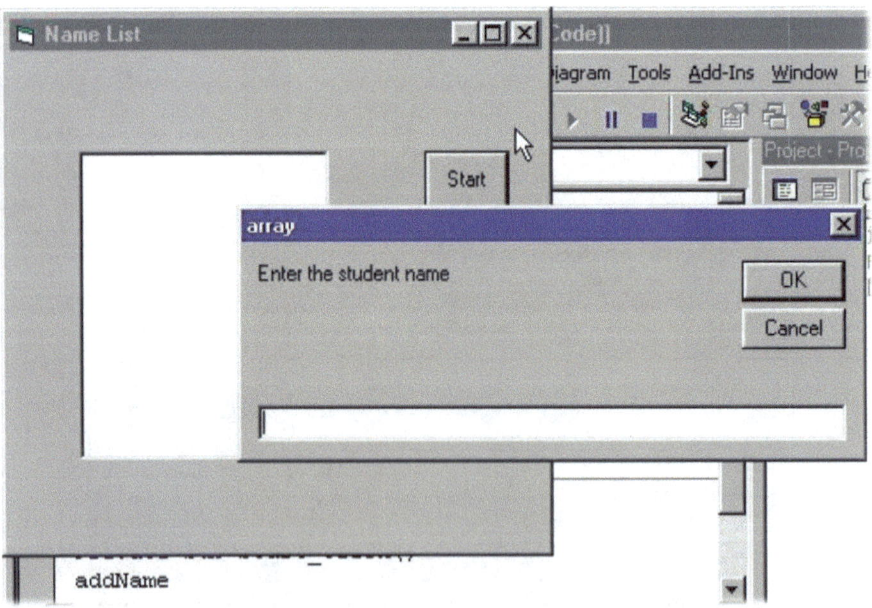

Msgboxes

Message boxes are used when we want to ask the user a question or display an error message(s) and advise the user. There are six types of message boxes. The list of all the possible msgbox events

Button Layout	Value	Short Description
vbOKonly	0	Displays the OK button.
vbOKCancel	1	Displays the Ok and cancel button.
vbAbortRetryIgnore	2	Displays the Abort, Retry and Ignore
vbYesNoCancel	3	Displays Yes, No and Cancel button
vbYesNo	4	Displays the Yes / No button
vbRetryCancel	5	Displays the retry and Cancel buttons.

Msgbox Return Value	Short Description
vbOk	The User Clicked OK
vbCancel	The User Clicked Cancel
vbAbort	The User Clicked Abort
vbRetry	The User Clicked Retry
vbIgnore	The User Clicked Ignore
VbYes	The User Clicked Yes
VbNo	The User Clicked No

The syntax for use of the message box in **Mgsgbox "TEXT", VALUE, "TITLE"** If you want to use two or more tables just add the values together. Therefore to print a message box to say "The Device was not Found!" OK & Explanation.

Mathematical Functions

The mathematical functions are very useful and important in programming. The common mathematical functions in Visual Basic are **Rnd, Sqr, Int, Abs, Exp, Log, Sin, Cos, Tan , Atn, Fix** and **Round.**

The functions are **Int, Sqr, Abs, Exp, Fix, Round** and **Log.**

a. *Int* is the function that converts a number into an integer by truncating its decimal part and the resulting integer is the largest integer that is smaller than the number. For example, Int(2.4)=2, Int(4.8)=4, Int(-4.6)= -5, Int(0.032)=0.

b. *Sqr* is the function that computes the square root of a number. For example, Sqr(4)=2, Sqr(9)=2.

c. *Abs* is the function that returns the absolute value of a number. So Abs(-9) = 9 and Abs(8)= 8.

d. *Exp* of a number x is the value of e^x. For example, Exp(1)=e^1 = 2.7182818284590

e. *Fix and Int* are the same if the number is a positive number as both truncate the decimal part of the number and return an integer.

   However, when the number is negative, it will return the smallest integer that is larger than the number. For example, Fix(-6.34)= -6 while Int(-6.34)=-7.

f. *Round* is the function that rounds up a number to a certain number of decimal places. The Format is Round (n, m) which means to round a number n to m decimal places. For example, Round (7.2567, 2) =7.26

g. *Log* is the function that returns the natural Logarithm of a number. For example, Log 10= 2.302585

Data Access

Data Access refers to the way a programmer interacts with the back end or the DBMS where most of the data are stored.

DAO (Data Access Objects)
RDO (Remote Data Objects)
ADO (Active Data objects)

Data Access Objects (DAO)

Data Access Object were an object oriented interface created by Microsoft which allowed early versions of Microsoft Access and Visual Basic to use the Jet database engine. DAO works by creating a "Workspace" object in which all database operations are performed. The workspace object exists as a session object that exists within a larger database engine object. There are two types of database engines:

- Jet database engine object,
- ODBCDirect database engine.

Microsoft Jet Database Engine

The Jet database engine object consists of several objects: a workspace object and a series of error objects. Recordset objects which are defined by a set of field objects, relation objects which show the relationship between different fields in the database and table definition objects which consists of fields and indexes of selected fields.

Microsoft Jet Database Engine is a database engine on which several Microsoft products were built. A database engine is the underlying component of a database, a collection of information stored on a computer in a systematic way.

ODBCDirect Database Engine

The ODBCDirect database engine consists of a workspace object and an errors object. The main difference between this database engine and the Jet database engine is that the workspace object is different in that it only consists of a series of ODBC connection objects and the database object consists of a series of recordset objects. The ODBC connection objects consist of QueryDef objects and recordset objects

Remote Data Objects

Remote Data Objects (abbreviated **RDO**) is the name of an obsolete data access application programming interface. This includes database connection, queries, stored procedures, result manipulation and change commits. It allowed developers to create interfaces that can directly interact with Open Database Connectivity (ODBC) data sources on remote machines, without having to deal with the comparatively complex ODBC API.

Microsoft ActiveX Data Objects

Microsoft **ActiveX Data Objects** (ADO) is a set of Component Object Model objects for accessing data sources. It provides a layer between programming languages and OLE DB, which allows a developer to write programs which access data, without knowing how the database is implemented. No knowledge of SQL is required to access a database when using ADO, although one can use ADO to execute arbitrary SQL commands. The disadvantage of this (i.e. using SQL directly) is that it introduces a dependency upon the type of database used.

Data Control

Data Control is the object which links a form to a database and allows you to access the fields in the tables making up the database. It's called Data in the Toolbox.

Visual Basic provides other objects that allow to link to databases. ADO (ActiveX Data Objects) is certainly more powerful and more efficient than the Data Control.

The arrow buttons on the control are used to navigate through the database records:

First record and **Previous**

Next and **Last record**

The buttons correspond to 4 methods. They are:

- MoveFirst
- MovePrevious
- MoveNext
- MoveLast

Properties of the Data Control

- *Name*: the name to use in code - Data1 is default we can have several data controls on the form.
- *Connect*: the kind of database. We can connect to Access, Foxpro, dBaseIV, etc.
- *DatabaseName*: the name and path of the database the control is connected to.
- *RecordSource*: the name of the database table being used.
- *BOFAction and EOFAction*: action to take when trying to read before the beginning of file or past the end of file.
- *Recordset*: this is a run time property, and it's an important one. It represents the result of the query executed to the database. It contains all the records required by this Data Control.Navigation through the database, navigate through the recordset, which is then mapped back to the database.

To bind the control to the database field we use its Data properties. They are

- **DataSource** is the name of the Data Control. Data Control specifies the name of the database to use the name of the table to access.

- DataField is the name of the field to bind. Field is selected from the content of the table.

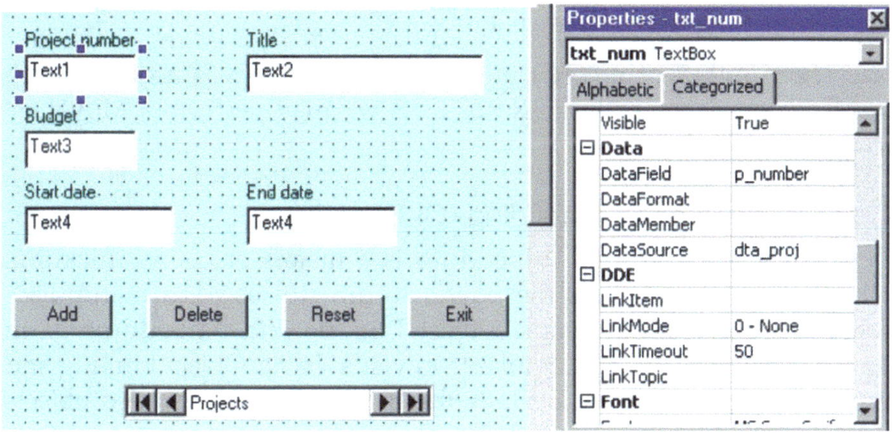

RecordSet Methods

Data1.RecordSet.AddNew

adds a new record to the recordset

Data1.RecordSet.Delete

deletes the current record

Data1.RecordSet.MoveNext

moves to the next record in the recordset

Data1.RecordSet.MovePrevious

moves to the previous record

Data1.RecordSet.MoveFirst

moves to the first record

Data1.RecordSet.MoveLast

moves to the last record

Data1.UpdateRecord

saves a changed record. This is also done automatically when we use any of the Move methods.

RecordSet Properties

Data1.RecordSet.EOF

True if MoveNext has moved us past the last record

Data1.RecordSet.BOF

True if MovePrevious has moved us in front of the first record

Data1.RecordSet.RecordCount

The number of records in the RecordSet. This is accurate with small RecordSets. For large RecordSets we may perform a MoveLast to get actual value.

Object Linking and Embedding (OLE)

Object Linking and Embedding is a means of communication, which gives applications the power to directly use and manipulate other windows applications. Object Linking and Embedding is an important windows topic available in visual basic. Object Linking and Embedding specification permits the user to link and embed objects and also edit the object within the container application.

Object Linking and Embedding is a technology that enables the programmer of windows based application to create an application, display data from many different applications and enables the user to edit that data from within the application in which it was created.

Creating Objects at Runtime.

To create a linked or embedded object at runtime, various methods and properties are used in code file called "ole.xls"

OLE1. createlink "c:\Ole.xls"

Creating an empty embedded object at run time

The create embed method can be used to specify an empty embedded object at ryn time. Microsoft Excel Worksheet in Ole container Control

OLE1.CreateEmbed ","Excel sheet".

Exercise to Practice

1. Define form
2. Explain IDE.
3. How to create menus at run time in VB?
4. ActiveX and Types of ActiveX Components in VB
5. Does VB Supports OOPS Concepts? Explain.

6. Advantage of ActiveX Dll over ActiveX Exe .
7. Can database schema be changed with DAO, RDO or ADO?
8. Can you create a tabletype of recordset in Jet-connected ODBC database engine?
9. Constructors and destructors.
10. Controls which do not have events.
11. Default property of datacontrol.
12. Define the scope of Public, Private, Friend procedures?
13. What are the ADO objects? Explain them.
14. What is OLE? Explain.
15. What is the difference between Msgbox Statement and Msgbox ( ) function?

CHAPTER 12

Database Management System (DBMS)

Database Management System

A collection of programs that enables us to store, modify, and extract information from a database. Database Management System is a database program that uses a standard method of cataloging, retrieving, and running queries on data. The Database Management System manages incoming data, organizes and provides ways for the data to be modified or extracted by users or other programs.

Database

A database is a collection of information that is organized so that it can easily be accessed, managed, and updated. In one view, databases can be classified according to types of content: bibliographic, full-text, numeric, and images. Databases are organized by fields, records, and files. A field is a single piece of information; a record is one complete set of fields and a file is a collection of records. For example, a telephone book is analogous to a file. It contains a list of records, each of which consists of three fields: name, address, and telephone number.

Database can be defined as a structured collection of records or data that is stored in a computer so that a program can consult it to answer queries. The records retrieved in answer to queries become information that can be used to make decisions. The computer program used to manage and query a database is known as a database management system (DBMS).

Data Model

A Data model instance may be one of three kinds

- a conceptual schema (data model) describes the semantics of an organization. This consists of entity classes (representing things of significance to the organization) and relationships (assertions about associations between pairs of entity classes).
- a logical schema (data model) describes the semantics, as represented by a particular data manipulation technology. This consists of descriptions of tables and columns.
- a physical schema (data model) describes the physical means by which data are stored. This is concerned with partitions, CPUs, tablespaces, and the like.

Relational Database Model

A relational database is a database that conforms to the relational model, and refers to a database's data and schema. Common usage of the term "Relational database management system" technically refers to the software used to create a relational database, but sometimes mistakenly refers to a relational database. A relational database is a database that conforms to the relational model, and could also be defined as a set of relations or a database built in an RDBMS.

A relational database management system (RDBMS) is a system that manages data using the relational model. Frequently, the term "RDBMS" is inaccurately used as a generic label for the relational database concept. Ironically, most RDBMS software packages are not technically considered "relational" because they do not fully conform to the relational model.

Table

A table is a set of data elements (values) that is organized using a model of horizontal rows and vertical columns. The columns are identified by name, and the rows are identified by the values appearing in a particular column subset which has been identified as a candidate key. Table is another term for relations; although there is the difference in that a table is usually a multi-set (bag) of rows whereas a relation is a set and does not allow duplicates. A table has a specified number of columns but can have any number of rows.

View

A view is a virtual or logical table composed of the result set of a query. Unlike ordinary tables (base tables) in a relational database, a view is not part of the physical schema it is a dynamic, virtual table computed or collated from data in the database.

Stored Procedure

A stored procedure is a subroutine available to applications accessing a relational database system. Stored procedures are actually stored in the database. Typical uses for stored procedures include data validation (integrated into the database) or access control mechanisms. Furthermore, stored procedures are used to consolidate and centralize logic that was originally implemented in applications. Stored procedures are similar to user-defined functions.

Logical data design

Logical data design concerns normalization of data followed by the creation of tables, views, indexes, and authorization schemes.

Physical data design

Physical specifying how database records are stored, accessed and related to ensure adequate performance.

Physical database design from three aspects

- what techniques for storing and finding data exist,
- which are implemented within a particular DBMS,
- which might be selected by the designer for a given application, knowing the properties of his data.

Reports

Reports are the formatted and organized presentation of data. Most database management systems include a report writer that enables you to design and generate reports.

Data Administration

The process of managing the institutional data in order to provide reliable, accurate, secure and accessible data to meet strategic and management needs at all levels of the enterprise. It is the purpose of this process to improve the accuracy, reliability, and security of data, reduce data redundancy; provide ease of access, assuring that data are easily located, accessible once located, and clearly defined and to provide data standards.

Database Administration

Database administration is the process of establishing computerised databases, and insuring their recoverability, integrity, security, availability, reliability, and performance. A database administrator (DBA) is a person who is responsible for the environmental aspects of a database. In general, these include

- Recoverability - Creating and testing Backups
- Integrity - Verifying or helping to verify data integrity
- Security - Defining and/or implementing access controls to the data
- Availability - Ensuring maximum uptime
- Performance - Ensuring maximum performance given budgetary constraints

Data Dictionary

A data dictionary is a set of metadata that contains definitions and representations of data elements. Within the context of a DBMS, a data dictionary is a read-only set of tables and views.

Data Definition Language Compiler

The DDL compiler converts the data definition statements into a set of tables. These tables contain the metadata concerning the database and are in a form that can be used by the other components of the DBMS.

Data Manager

The data manager is the control software components of the DBMS. One of the functions of the data manager is to convert operations in the user's queries coming directly via the query processor or indirectly via an application program from the user's logical view to the physical system.

File Manager

Responsibility for the structure of the files and managing the file spaces rests with the file managers.

Disk Manager

The disk manager transfer the blocks or page requested by the file manager so that the latter need not be concerned with the physical characteristics of the underlying storage media.

Query Processor

The query processor is used to interpret the on-line user's query and convert into an efficient series of operations in a form capable of being sent to the manager for execution.

Data Files

Data files contain the data portion of the database.

Access Aids

To improve the performance of a DBMS a set of access aid in forms of indexes are usually provided in a database system.

Concurrency

Concurrency is a property of systems in which several computational processes are executing at the same time, and potentially interacting with each other. The study of concurrency encompasses a broad range of systems, from tightly-coupled, largely synchronous parallel computing systems to loosely-coupled, largely asynchronous distributed systems.

The concurrent processes may be executing truly simultaneously, in the case that they run on separate processors, or their execution steps may be interleaved to produce the appearance of concurrency, as in the case of separate processes running on a multitasking system. Because the processes in a concurrent system can interact with each other while they are executing, the number of possible execution paths in the system can be extremely large and the resulting behavior can be very complex.

The difficulties associated with concurrency have been tackled both through the construction of languages and concepts to make the complexity of concurrent execution manageable, and through the development of theories for reasoning about interacting concurrent processes.

Data Warehouse

A data warehouse is the main repository of an organization's historical data, its corporate memory. It contains the raw material for management's decision support system. The critical factor leading to the use of a data warehouse is that a data analyst can perform complex queries and analysis such as data mining on the information without slowing down the operational systems. Data warehouse in the following terms; it is

- subject-oriented, meaning that the data in the database is organized so that all the data elements relating to the same real-world event or object are linked together;
- time-variant, meaning that the changes to the data in the database are tracked and recorded so that reports can be produced showing changes over time;
- non-volatile, meaning that data in the database is never over-written or deleted, once committed, the data is static, read-only, but retained for future reporting; and
- integrated, meaning that the database contains data from most or all of an organization's operational applications, and that this data is made consistent.

Database Software

Access

Microsoft Access, is a relational database management system from Microsoft which combines the relational Microsoft Jet Database Engine with a graphical user interface. Access can use data stored in Access/Jet, Microsoft SQL Server, Oracle or any ODBC-compliant data container.

Oracle

An **Oracle database** consists of a collection of data managed by an Oracle database management system. Popular usage also uses the term to refer to the Oracle DBMS software, but not necessarily to a specific database under its control.

SQL Server

SQL Server was designed for client/server use and is accessed by applications using SQL. It runs on OS/2, Windows NT, NetWare servers, VAXen, and UNIX workstations. Generically, any database management system (DBMS) that can respond to queries from client machines formatted in the SQL language.

MySql

MySQL is an open source RDBMS that relies on SQL for processing the data in the database. MySQL provides APIs for the languages C, C++, Eiffel, Java, Perl, PHP and Python. MySQL is most commonly used for Web applications, embedded applications and has become a popular alternative to proprietary database systems because of its speed and reliability.

Components of a DBMS

A DBMS is a complex software system that is used to manage, store and manipulate data and the metadata used to describe the data.

DBMS Facility

Data Definition Language

The definition may also include any constraints that have to be maintained. These definitions which can be described as metadata about the data in the database are expressed in the DDL of the DBMS and maintained in compiled form (usually as a set of tables).

Data Manipulation Language

The language used to manipulate data in the database is called Data Manipulation Language (DML). This involves retrieval of data into the database and deletion or modification of existing data. The first of these data manipulation operations is called Query.

The subset of the DML, used to pose a query is known as a Query Language. The main component of a database management system are database users, database administrator and application programs.

Any person who can access the database directly using a DBMS is called database user.

Database Administrator

The DBA must be a manager, more than a technician- seeking to meet the needs of people who use data. Since many users may share the same data resources, the DBA must be prepared to mediate conflicting needs and objectives.

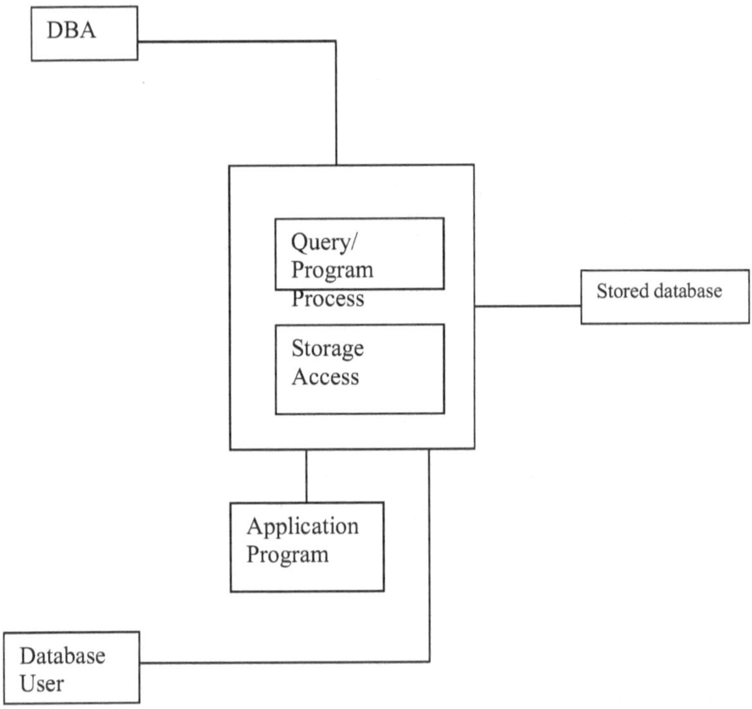

Components of DBMS

Architecture for a DBMS

The architecture is divided into three general levels.

a) *External level (physical level)* - It is the one closet to the physical storage (ie) this is the one concerned with the way in which the data is actually stored.

b) *Conceptual level (view level)* - It is the one closet to the users (ie) this is the one concerned with the way in which the data is viewed by individual users.

c) *Internal level*: This is the lowest level of abstraction. It is closest to hardware internal view describes how the data are physically stored. It also describes the data structure and access method to be used by the database. Internal view is closest to the physical storage.

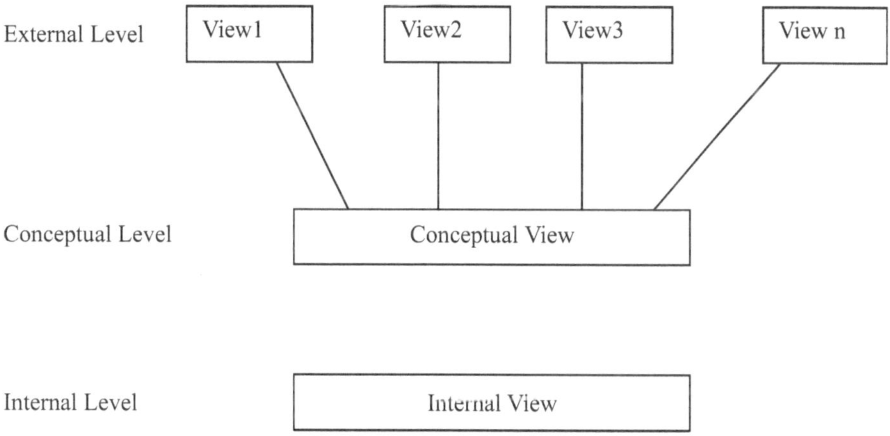

Three levels of DBMS

The DBMS accepts SQL commands generated from a variety of user interfaces, produces query evaluation plans, executes these plans against the database, and returns the answers.

When a user issues a query, the parsed query is presented to a query optimizer, which uses information about how the data is stored to produce an efficient execution plan for evaluating the query. An execution plan is a blueprint for evaluating a query, and is usually represented as a tree of relational operators. Relational operators serve as the building blocks for evaluating queries posed against the data.

The code that implements relational operators sits on top of the file and access methods layer. This layer includes a variety of software for supporting the concept of a file, which, in a DBMS, is a collection of pages or a collection of records. This layer typically supports a heap file, or file of unordered pages, as well as indexes. In addition to keeping track of the pages in a file, this layer organizes the information within a page. The files and access methods layer code sits on top of the buffer manager, which brings pages in from disk to main memory as needed in response to read requests.

The lowest layer of the DBMS software deals with management of space on disk, where the data is stored. Higher layers allocate, deallocate, read, and write pages through (routines provided by) this layer, called the disk space manager.

The DBMS supports concurrency and crash recovery by carefully scheduling user requests and maintaining a log of all changes to the database. DBMS components associated with concurrency control and recovery include the transaction manager, which ensures that transactions request and release locks according to a suitable locking protocol and schedules the execution transactions; the lock manager, which keeps track of requests for locks and grants locks on database objects when they become available; and the recovery manager, which is responsible for maintaining a log and restoring the system to a consistent state after a crash. The disk space manager, buffer manager, and file and access method layers must interact with these components as shown in the following diagram:

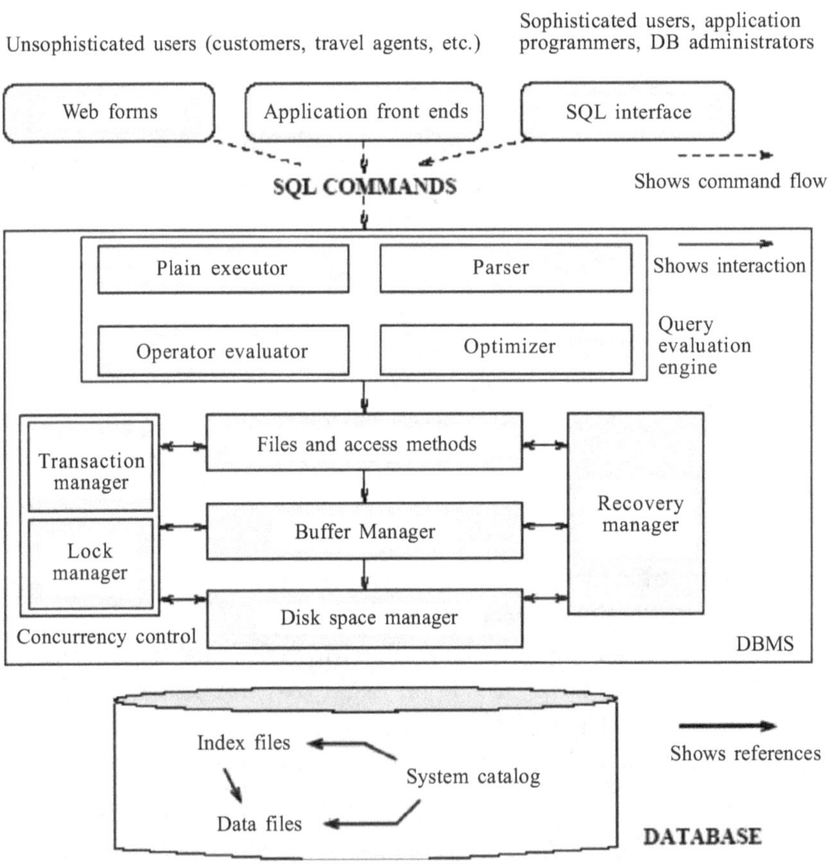

Architecture of a DBMS

Data Modeling

This is a form of modeling that applies the Relational Data Model to data storage. The principle diagram is an Entity Relationship Diagram. In data modelling, we are structuring and organizing data. These data structures are then typically implemented in a database management system. In addition to defining and organizing the data, data modeling will impose (implicitly or explicitly) constraints or limitations on the data placed within the structure.

In computer science, data modeling is the process of structuring and organizing data. These data structures are then typically implemented in a database management system. In addition to defining and organizing the data, data modeling may also impose constraints or limitations on the data placed within the structure.

Managing large quantities of structured and unstructured data is a primary function of information systems. Data models describe structured data for storage in data management systems such as relational databases. They typically do not describe unstructured data, such as word processing documents, email messages, pictures, digital audio, and video. Early phases of many software development projects emphasize the design of a conceptual data model. Such a design can be detailed into a logical data model. In later stages, this model may be translated into physical data model.

A data model is a conceptual representation of the data structures that are required by a database. The data structures include the data objects, the associations between data objects, and the rules which govern operations on the objects. As the name implies, the data model focuses on what data is required and how it should be organized rather than what operations will be performed on the data. To use a common analogy, the data model is equivalent to an architect's building plans.

A data model is independent of hardware or software constraints. Rather than try to represent the data as a database would see it, the data model focuses on representing the data as the user sees it in the "real world". It serves as a bridge between the concepts that make up real-world events and processes and the physical representation of those concepts in a database.

Methodology

There are two major methodologies used to create a data model: the Entity-Relationship (ER) approach and the Object Model. This document uses the Entity-Relationship approach.

Data Modeling In the Context of Database Design

Database design is defined as: "design the logical and physical structure of one or more databases to accommodate the information needs of the users in an organization for a defined set of applications". The design process roughly follows five steps:

1. planning and analysis
2. conceptual design
3. logical design
4. physical design
5. implementation

The data model is one part of the conceptual design process. The other, typically is the functional model. The data model focuses on what data should be stored in the database while the functional model deals with how the data is processed. To put this in the context of the relational database, the data model is used to design the relational tables. The functional model is used to design the queries which will access and perform operations on those tables.

Components of a Data Model

The data model gets its inputs from the planning and analysis stage. Here the modeler, along with analysts, collects information about the requirements of the database by reviewing existing documentation and interviewing end-users.

The data model has two outputs. The first is an entity-relationship diagram which represents the data structures in a pictorial form. Because the diagram is easily learned, it is valuable tool to communicate the model to the end-user. The second component is a data document. This document describes in detail the data objects, relationships, and rules required by the database. The dictionary provides the detail required by the database developer to construct the physical database.

Importance of Data Modeling

The goal of the data model is to make sure that the all data objects required by the database are completely and accurately represented. Because the data model uses easily understood notations and natural language, it can be reviewed and verified as correct by the end-users.

The data model is also detailed enough to be used by the database developers to use as a "blueprint" for building the physical database.

The information contained in the data model will be used to define the relational tables, primary and foreign keys, stored procedures, and triggers. A poorly designed database will require more time in the long-term. Without careful planning we may create a database that omits data required to create critical reports, produces results that are incorrect or inconsistent, and is unable to accommodate changes in the user's requirements.

RDBMS (Relational Database Management System)

Relational Database Management System and pronounced as separate letters, a type of database management system (DBMS) that stores data in the form of related tables. Relational databases are powerful because they require few assumptions about how data is related or how it will be extracted from the database. As a result, the same database can be viewed in many different ways.

An important feature of relational systems is that a single database can be spread across several tables. This differs from flat-file databases, in which each database is self-contained in a single table. Almost all full-scale database systems are RDBMS's. Small database systems, however, use other designs that provide less flexibility in posing queries.

Client/Server Model of Computing

The client/server model is a form of distributed computing where one program (the client) communicates with another program (the server) for the purpose of exchanging information.

The client's responsibility is usually to:

1. Handle the user interface.
2. Translate the user's request into the desired protocol.
3. Send the request to the server.
4. Wait for the server's response.
5. Translate the response into "human-readable" results.
6. Present the results to the user.

The server's functions include:

1. Listen for a client's query.
2. Process that query.
3. Return the results back to the client.

A typical client/server interaction goes like this:

1. The user runs client software to create a query.
2. The client connects to the server.
3. The client sends the query to the server.
4. The server analyzes the query.
5. The server computes the results of the query.
6. The server sends the results to the client.
7. The client presents the results to the user.
8. Repeat as necessary.

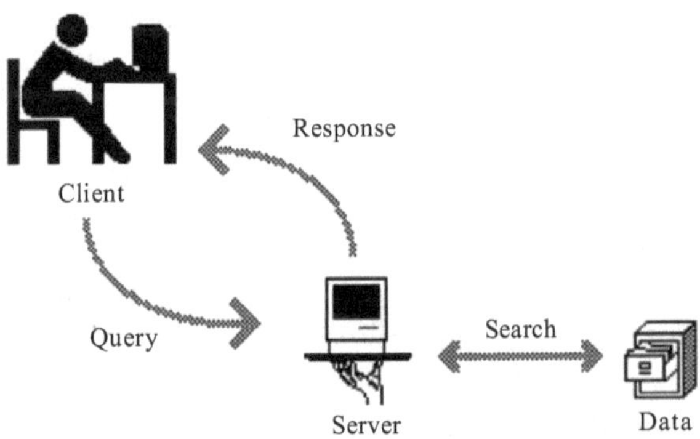

A typical client/server interaction

This client/server interaction is a lot like going to a French restaurant. At the restaurant, you (the user) are presented with a menu of choices by the waiter (the client). After making your selections, the waiter takes note of our choices, translates them into French and presents them to the French chef (the server) in the kitchen. After the chef prepares your meal, the waiter returns with our diner (the results). Hopefully, the waiter returns with the items you selected, but not always; sometimes things get "lost in the translation."

Flexible user interface development is the most obvious advantage of client/server computing. It is possible to create an interface that is independent of the server hosting the data. Therefore, the user interface of a client/server application can be written on a Macintosh and the server can be written on a mainframe. Clients could be also written for DOS- or UNIX-based computers. This allows information to be stored in a central server and disseminated to different types of remote computers. Since the user interface is the responsibility of the client, the server has more computing resources to spend on analyzing queries and disseminating information.

This is another major advantage of client/server computing; it tends to use the strengths of divergent computing platforms to create more powerful applications. Although, its computing and storage capabilities are dwarfed by those of the mainframe, there is no reason why a Macintosh could not be used as a server for less demanding applications.

In short, client/server computing provides a mechanism for disparate computers to cooperate on a single computing task.

Client/server is network architecture which separates a client (often an application that uses a graphical user interface) from a server. Each instance of the client software can send requests to a *server*. Specific Types of servers include: application servers, file servers, terminal servers, and mail servers. While their purpose varies somewhat, the basic architecture remains the same.

Characteristics of a server:

- Passive (slave)
- Waits for requests
- Upon receipt of requests, processes them and then serves replies

Characteristics of a client:

- Active (master)
- Sends requests
- Waits for and receives server replies

Servers can be stateless or stateful. A stateless server does not keep any information between requests. A stateful server can remember information between requests. The scope of this information can be global or session. A HTTP server for static HTML pages is an example of a stateless server while Apache Tomcat is an example of a stateful server. The interaction between client and server is often described using sequence diagrams. Sequence diagrams are standardized in the UML.

Another type of network architecture is known as a peer-to-peer architecture because each node or instance of the program is both a "client" and a "server" and each has equivalent responsibilities. Both architectures are in wide use.

Entity Relationship Diagram (E-R diagram)

A type of diagram used in data modeling for relational data bases. These diagrams show the structure of each table and the links between tables.

Its components are

- Rectangles representing entity sets.
- Ellipses representing attributes.
- Diamonds representing relationship sets.
- Lines linking attributes to entity sets and entity sets to relationship sets.

Entity Relationship Diagrams (ERDs) illustrate the logical structure of databases.

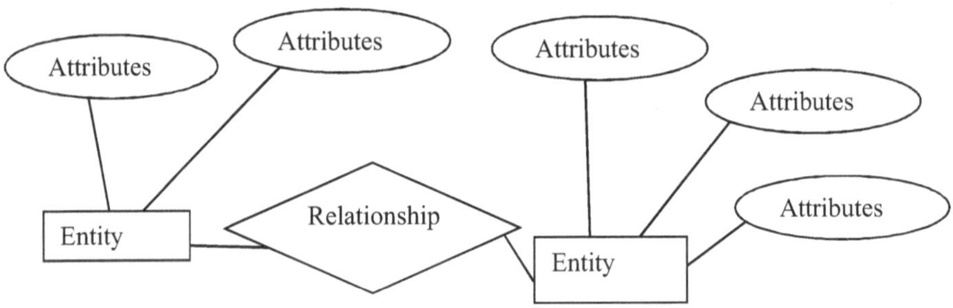

An ER Diagram

Entity Relationship Diagram Notations

Entity

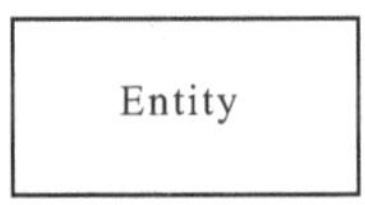

An entity is an object or concept about which you want to store information. Entities are usually recognizable concepts, either concrete or abstract, such as person, places, things, or events which have relevance to the database. Some specific examples of entities are EMPLOYEES, PROJECTS, INVOICES.

Attribute

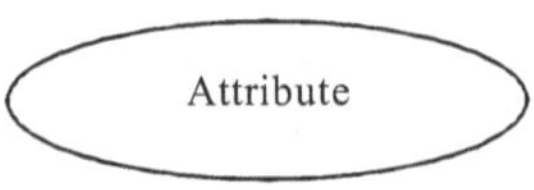

A key attribute is the unique, distinguishing characteristic of the entity. For example, an employee's social security number might be the employee's key attribute.

Relationships

Relationships illustrate how two entities share information in the database structure.

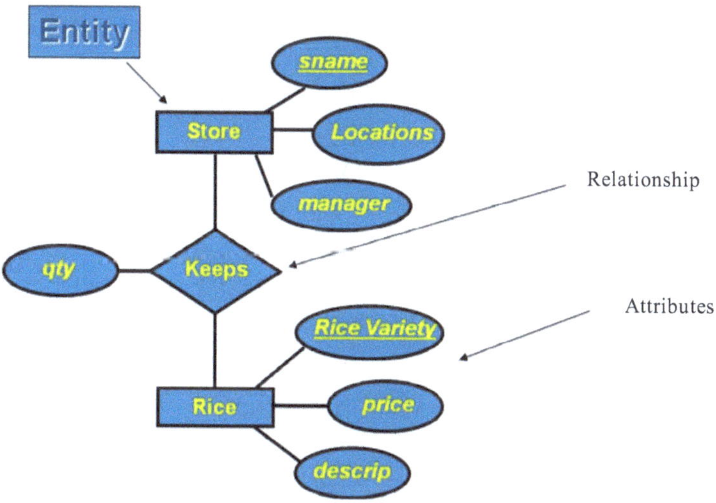

Example for E-R diagram

Front-end and Back-end

1. For software applications, front end is the same as user interface.
2. In client/server applications, the client part of the program is often called the front end and the server part is called the back end.

In their most general meanings, the terms front-end and back-end refer to the initial and the end stages of a process flow. These terms acquire more special meanings in particular areas. The general idea is that the front-end is responsible for collecting input from the user, which can be in a variety of forms, and processing it in such a way that it conforms to a specification that the back-end can use. The connection of the front-end to the back-end is a kind of interface.In software design, the front-end is the part of a software system that interacts directly with the user, and the back-end comprises the components that process the output from the front-end.

Codd's Rules

Rule 1 : The information Rule.

"All information in a relational data base is represented explicitly at the logical level and in exactly one way - by values in tables." Everything within the database exists in tables and is accessed via table access routines.

Rule 2 : Guaranteed access Rule.

"Each and every datum (atomic value) in a relational data base is guaranteed to be logically accessible by resorting to a combination of table name, primary key value and column name."

Rule 3 : Systematic treatment of null values.

"Null values (distinct from the empty character string or a string of blank characters and distinct from zero or any other number) are supported in fully relational DBMS for representing missing information and inapplicable information in a systematic way, independent of data type."

If data does not exist or does not apply then a value of NULL is applied, this is understood by the RDBMS as meaning non-applicable data.

Rule 4 : Dynamic on-line catalog based on the relational model

"The data base description is represented at the logical level in the same way as-ordinary data, so that authorized users can apply the same relational language to its interrogation as they apply to the regular data."

The Data Dictionary is held within the RDBMS, thus there is no-need for off-line volumes to tell you the structure of the database.

Rule 5 : Comprehensive data sub-language Rule.

"A relational system may support several languages and various modes of terminal use (for example, the fill-in-the-blanks mode). However, there must be at least one language whose statements are expressible, per some well-defined syntax, as character strings and that is comprehensive in supporting all the following items

- Data Definition
- View Definition

- Data Manipulation (Interactive and by program).
- Integrity Constraints
- Authorization.

Every RDBMS should provide a language to allow the user to query the contents of the RDBMS and also manipulate the contents of the RDBMS.

Rule 6 : .View updating Rule

"All views that are theoretically updatable are also updatable by the system."

Not only can the user modify data, but so can the RDBMS when the user is not logged-in.

Rule 7 : High-level insert, update and delete.

"The capability of handling a base relation or a derived relation as a single operand applies not only to the retrieval of data but also to the insertion, update and deletion of data."

The user should be able to modify several tables by modifying the view to which they act as base tables.

Rule 8 : Physical data independence.

"Application programs and terminal activities remain logically unimpaired whenever any changes are made in either storage representations or access methods."

The user should not be aware of where or upon which media data-files are stored

Rule 9 : Logical data independence.

"Application programs and terminal activities remain logically unimpaired when information-preserving changes of any kind that theoretically permit un-impairment are made to the base tables."

User programs and the user should not be aware of any changes to the structure of the tables (such as the addition of extra columns).

Rule 10 : Integrity independence.

"Integrity constraints specific to a particular relational data base must be definable in the relational data sub-language and storable in the catalog, not in the application programs."

If a column only accepts certain values, then it is the RDBMS which enforces these constraints and not the user program, this means that an invalid value can never be entered into this column, whilst if the constraints were enforced via programs there is always a chance that a buggy program might allow incorrect values into the system.

Rule 11 : Distribution independence.

"A relational DBMS has distribution independence."

The RDBMS may spread across more than one system and across several networks, however to the end-user the tables should appear no different to those that are local.

Rule 12 : Non-subversion Rule.

"If a relational system has a low-level (single-record-at-a-time) language, that low level cannot be used to subvert or bypass the integrity Rules and constraints expressed in the higher level relational language (multiple-records-at-a-time)."

SQL – Structured Query Language

SQL stands for Structured Query Language. SQL helps to access and manipulate databases.

SQL is a non-procedural database language that provides facilities for:

- Execute queries against a database
- Retrieve data from a database
- Insert records in a database
- Update records in a database
- Delete records from a database
- Create new databases
- Create new tables in a database
- Create stored procedures in a database
- Create views in a database
- Handles security and privacy constraints.

SQL DML and DDL

SQL can be divided into two parts:

- Data Definition Language (DDL).
- Data Manipulation Language (DML)

Data Definition Language (DDL)

Data Definition Language statements are used to define the database structure or schema. Data Definition Language statements are used to build and modify the structure of your tables and other objects in the database.

The most important Data Definition Language statements in SQL are

- CREATE DATABASE - creates a new database
- ALTER DATABASE - modifies a database
- CREATE TABLE - creates a new table
- ALTER TABLE - modifies a table
- DROP TABLE - deletes a table
- CREATE INDEX - creates an index (search key)
- DROP INDEX - deletes an index

Data Manipulation Language (DML)

Data Manipulation Language statements are used for managing data within schema objects. Data Manipulation Language statements are used to work with the data in tables. When you are connected to most multi-user databases.

The query and update commands form the Data Manipulation Language are

- SELECT - extracts data from a database
- UPDATE - updates data in a database
- DELETE - deletes data from a database
- INSERT INTO - inserts new data into a database

The DDL part of SQL permits database tables to be created or deleted. It also defines indexes (keys), specify links between tables, and impose constraints between tables.

CREATE TABLE

The CREATE TABLE statement is used to create a table in a database.

CREATE TABLE Syntax

CREATE TABLE table_name

```
(
column_name1 data_type,
column_name2 data_type,
column_name3 data_type,
....
)
```

INSERT TABLE

The INSERT INTO statement is used to insert new records in a table.

INSERT INTO Syntax

It is possible to write the INSERT INTO statement in two forms. The first form doesn't specify the column names where the data will be inserted, only their values:

```
INSERT INTO table_name
VALUES (value1, value2, value3,...)
```

The second form specifies both the column names and the values to be inserted:

```
INSERT INTO table_name (column1, column2, column3,...)
VALUES (value1, value2, value3,...)
```

SELECT Statement

The SELECT statement is used to select data from a database. The statement begins with the SELECT keyword. The basic SELECT statement has 3 clauses:

- SELECT
- FROM
- WHERE

The SELECT clause specifies the table columns that are retrieved. The FROM clause specifies the tables accessed. The WHERE clause specifies which table rows are used. The WHERE clause is optional.

SELECT Syntax

```
SELECT column_name(s)
FROM table_name
```

and

```
SELECT * FROM table_name
```

WHERE Clause

The WHERE clause is used to filter records. The WHERE clause is used to extract only those records that fulfill a specified criterion.

WHERE Syntax

```
SELECT column_name(s)
FROM table_name
WHERE column_name operator value
```

ORDER BY Keyword

The ORDER BY keyword is used to sort the result-set by a specified column. The ORDER BY keyword sort the records in ascending order by default. If we want to sort the records in a descending order, we can use the DESC keyword.

ORDER BY Syntax

```
SELECT column_name(s)
FROM table_name
ORDER BY column_name(s) ASC|DESC
```

The UPDATE Statement

The UPDATE statement is used to update existing records in a table.

UPDATE Syntax

```
UPDATE table_name
SET column1=value, column2=value2,...
WHERE some_column=some_value
```

DELETE TABLE

The DELETE statement is used to delete records in a table.

DELETE Syntax

```
DELETE FROM table_name
WHERE some_column=some_value
```

DROP TABLE

The DROP TABLE statement is used to delete a table.

FUNCTIONS

AVG

The AVG() function returns the average value of a numeric column.

AVG( ) Syntax

SELECT AVG(column_name) FROM table_name

MIN

The MIN() function returns the smallest value of the selected column.

MIN( ) Syntax

SELECT MIN(column_name) FROM table_name

MAX

The MAX() function returns the largest value of the selected column.

MAX() Syntax

SELECT MAX(column_name) FROM table_name

COUNT

The COUNT() function returns the number of rows that matches a specified criteria.

COUNT(column_name) Syntax

The COUNT(column_name) function returns the number of values (NULL values will not be counted) of the specified column:

SELECT COUNT(column_name) FROM table_name

COUNT(*) Syntax

The COUNT(*) function returns the number of records in a table:

SELECT COUNT(*) FROM table_name

COUNT (DISTINCT column_name) Syntax

The COUNT (DISTINCT column_name) function returns the number of distinct values of the specified column:

SELECT COUNT(DISTINCT column_name) FROM table_name

SUM

The SUM() function returns the total sum of a numeric column.

SUM( ) Syntax

```
SELECT SUM(column_name) FROM table_name
```

GROUP BY Statement

The GROUP BY statement is used in conjunction with the aggregate functions to group the result-set by one or more columns.

GROUP BY Syntax

```
SELECT column_name, aggregate_function(column_name)
FROM table_name
WHERE column_name operator value
GROUP BY column_name
```

The HAVING Clause

The HAVING clause was added to SQL because the WHERE keyword could not be used with aggregate functions.

HAVING Syntax

```
SELECT column_name, aggregate_function(column_name)
FROM table_name
WHERE column_name operator value
GROUP BY column_name
HAVING aggregate_function(column_name) operator value
```

The UCASE() Function

The UCASE() function converts the value of a field to uppercase.

UCASE() Syntax

```
SELECT UCASE(column_name) FROM table_name
```

The LCASE() Function

The LCASE() function converts the value of a field to lowercase.

LCASE() Syntax

SELECT LCASE(column_name) FROM table_name

DCL – Data Control Language

DCL commands are used to enforce database security in a multiple user database environment. It is used to create roles, permissions, and referential integrity as well it is used to control access to database by securing it. Two types of DCL commands are GRANT and REVOKE. Only Database Administrator's or owner's of the database object can provide/remove privileges on a database object.

GRANT Command

SQL GRANT is a command used to provide access or privileges on the database objects to the users.

The Syntax for the GRANT command is:

```
GRANT privilege_name
ON object_name
TO {user_name |PUBLIC |role_name}
[WITH GRANT OPTION];
```

- privilege_name is the access right or privilege granted to the user. Some of the access rights are ALL, EXECUTE, and SELECT.
- object_name is the name of an database object like TABLE, VIEW, STORED PROC and SEQUENCE.
- user_name is the name of the user to whom an access right is being granted.
- PUBLIC is used to grant access rights to all users.
- ROLES are a set of privileges grouped together.
- WITH GRANT OPTION - allows a user to grant access rights to other users.

REVOKE Command

The REVOKE command removes user access rights or privileges to the database objects.

The Syntax for the REVOKE command is:

```
REVOKE privilege_name
```

ON object_name

FROM {user_name |PUBLIC |role_name}

Transaction Control (TCL)

Transaction Control (TCL) statements are used to manage the changes made by DML statements. It is used to manage different transactions occurring within a database. It allows statements to be grouped together into logical transactions.

- COMMIT - save work done
- SAVEPOINT - identify a point in a transaction to which you can later roll back
- ROLLBACK - restore database to original since the last COMMIT
- SET TRANSACTION - Change transaction options like isolation level and what rollback segment to use.

Privileges

Privileges define the access rights provided to a user on a database object. There are two types of privileges.

i. System privileges - This allows the user to CREATE, ALTER, or DROP database objects.

ii. Object privileges - This allows the user to EXECUTE, SELECT, INSERT, UPDATE, or DELETE data from database objects to which the privileges apply.

Example

To create table

create table labour (lname varchar(25),lid integer primary key,days integer ,netsal integer);

create table dept (lname varchar(25),lid integer primary key,dept varchar(20));

Insert the values into the table

insert into labour values('hari krishnan',5,30,2400);

Select the data from the table

select * from labour;

lname	lid	days	netsal
kumar	1	25	2000
ramakrishnan	2	30	2400
shiva	3	30	2400
priya	4	29	2320
hari krishnan	5	30	2400

select * from dept;

lname	lid	dept
kumar	1	fert
ramakrishnan	2	crop
shiva	3	crop
priya	4	sericulture
hari Krishnan	5	sericulture

select * from labour where netsal = 2400;

lname	lid	days	netsal
ramakrishnan	2	30	2400
shiva	3	30	2400
hari krishnan	5	30	2400

HAVING options

Table is created using create statement and the values are inserted using insert statement. Table looks as shown below

store_name	Sales	Date
Coimbatore	1500	Jan-05-1999
Erode	250	Jan-07-1999
Coimbatore	300	Jan-08-1999
Tirupur	700	Jan-08-1999

Query

SELECT store_name, SUM(sales)

FROM Store_Information

```
GROUP BY store_name
HAVING SUM(sales) > 1500
```

Output

store_name	SUM(Sales)
Coimbatore	1800

LIKE options

Query

Select store_name FROM store_Information WHERE store_name LIKE "E%";

Output

store_name	Sales	Date
Erode	250	Jan-07-1999

View command

Query

CREATE VIEW "VIEW_NAME" AS "SQL Statement"

Create the table Customer with(First_Name char(50), Last_Name char(50), Address char(50),City char(50), Country char(25), Birth_Date date) and create a view called V_Customer that contains only the First_Name, Last_Name, and Country columns from this table, we would type,

```
CREATE VIEW V_Customer
AS SELECT First_Name, Last_Name, Country
FROM Customer
```

Now we have a view called V_Customer with the following structure:

```
View V_Customer
(First_Name char(50),
Last_Name char(50),
Country char(25))
```

We can also use a view to apply joins to two tables. In this case, users only see one view rather than two tables. The two tables are given below

Table1 - Store_Information

store_name	Sales	Date
Coimbatore	1500	Jan-05-1999
Erode	250	Jan-07-1999
Coimbatore	300	Jan-08-1999
Tirupur	700	Jan-08-1999

Table2 - Geography

region_name	store_name
East	Tirupur
East	Salem
West	Coimbatore
West	Erode

We can build a view that has sales by region information. SQL statements are as follows:

```
CREATE VIEW V_REGION_SALES
AS SELECT A1.region_name REGION, SUM(A2.Sales) SALES
FROM Geography A1, Store_Information A2
WHERE A1.store_name = A2.store_name
GROUP BY A1.region_name
```

This gives us a view, V_REGION_SALES, that has been defined to store sales by region records. If we want to find out the content of this view, we type in

```
SELECT * FROM V_REGION_SALES
```

Output

REGION	SALES
East	700
West	2050

JOIN option

Query

Create the table with the following entity

create table labour (lname varchar(25),lid integer primary key,days integer ,netsal integer);

create table dept (lname varchar(25),lid integer primary key,dept varchar(20));

select labour.lname,dept.dept from labour JOIN dept ON (labour.lname = dept.dept);

Output

lname **dept**

lname from labour table and dept from dep table are joined.

Aggregate Functions

create table labour (lname varchar(25),lid integer primary key,days integer, netsal integer);

lname	lid	days	netsal
kumar	1	25	2000
ramakrishnan	2	30	2400
shiva	3	30	2400
priya	4	29	2320
hari krishnan	5	30	2400

Query

SELECT COUNT (*) FROM labour;

Output

COUNT

5

Query

SELECT SUM (netsal) FROM labour;

Output

SUM
11520

Query

SELECT MAX(netsal) FROM labour;

Output

MAX
2400

Query

SELECT MIN(netsal) FROM labour;

Output

MIN
2000

Query

SELECT AVG(netsal l) FROM labour;

Output

AVG
2304

Exercise to Practice

1. Define Data Base Management System?
2. List the applications of DBMS.
3. What are the advantages of using a DBMS?
4. Define instance and schema?
5. Define the terms 1) physical schema 2) logical schema.
6. What is conceptual schema?
7. Define data model?
8. What is a data dictionary?

9. What is an entity relationship model?
10. What are attributes? Give examples.
11. What is relationship? Give examples
12. Define the terms
    a. Entity set
    b. Relationship set
13. Define the terms
    i. DDL
    ii. DML
14. Write short notes on relational model
15. Define tuple and attribute
16. What is a SELECT operation?
17. Define query language?
18. What are the parts of SQL language?
19. What are the categories of SQL command?
20. What are the three classes of SQL expression?
21. Give the general form of SQL query?
22. What are aggregate functions? And list the aggregate functions supported by SQL?
23. What is view in SQL? How is it defined?
24. List the table modification commands in SQL?
25. List out the statements associated with a database transaction?
26. What is transaction?
27. Name the various privileges in SQL?
28. Mention the various user privileges.
29. Create the table EMP and DEPT and execute the following

Structure of EMP table and execute the following queries

Column name	Type	Description
EMPNO	NUMBER(4)	Employee Number
ENAME	VARCHAR(20)	Employee Name
JOB	CHAR(10)	Designation
MGR	NUMBER(4)	Respective Managers EMPNO
HIREDATE	DATE	Date of Joining
SAL	NUMBER(9,2)	Basic Salary
DEPTNO	NUMBER(2)	Department Number

Structure of DEPT Table

Column Names	Type	Description
DEPTNO	NUMBER(2)	Department Number
DNAME'	VARCHAR(20)	Department Name
LOC	VARCHAR(10)	Location of the Department

Query

```
Select * from emp WHERE deptno=20;
Select ename, sal FROM emp where sal>12500;
Select ename FROM emp where job = "AP" AND deptno = 20;
Select ename FROM emp where hiredate <= "30-SEP-91"
Select * from emp WHERE sal BETWEEN 10000 AND 12500;
Select ename FROM emp WHERE ename LIKE "s%"
```

30. Create the table sales with the following entity
    1. Sales number
    2. Product
    3. Amount
    4. Date

    Create the table customer with the following entity
    1. Customer number
    2. Customer name
    3. Product purchased
    4. Amount

    Use the VIEW and JOIN option.

31. Create the table EMP with the following entities and execute the queries

Column name	Type	Description
EMPNO	NUMBER(4)	Employee Number
ENAME	VARCHAR(20)	Employee Name
JOB	CHAR(10)	Designation
MGR	NUMBER(4)	Respective Managers EMPNO
HIREDATE	DATE	Date of Joining
SAL	NUMBER(9,2)	Basic Salary
DEPTNO	NUMBER(2)	Department Number

Query

```
Select COUNT (*) FROM emp;
Select SUM (sal) FROM emp;
Select MAX(sal) FROM emp;
Select MIN(sal) FROM emp;
Select AVG(sal) FROM emp;
```

32. Create the table EMP and DEPT with the following entities and execute the queries

Column name	Type	Description
EMPNO	NUMBER(4)	Employee Number
ENAME	VARCHAR(20)	Employee Name
JOB	CHAR(10)	Designation
MGR	NUMBER(4)	Respective Managers EMPNO
HIREDATE	DATE	Date of Joining
SAL	NUMBER(9,2)	Basic Salary
DEPTNO	NUMBER(2)	Department Number

DEPT Table

Column Names	Type	Description
DEPTNO	NUMBER(2)	Department Number
DNAME'	VARCHAR(20)	Department Name
LOC	VARCHAR(10)	Location of the Department

Query

Altering a table

ALTER TABLE emp ADD grade CHAR(2);

To delete table

DROP TABLE emp;

Nested Queries

SELECT deptno FROM emp WHERE ename = 'Arun';

SELECT ename FROM emp WHERE deptno = 10;

Combining the above two queries to get the results.

SELECT ename FROM emp WHERE deptno = (SELECT deptno FROM emp WHERE ename = 'Arun');

List the names of employee drawing the highest salary;

SELECT ename FROM emp WHERE sal = (SELECT MAX (sal) FROM emp);

List the names of the employees who got more than 1 increments

SELECT empno,ename FROM emp WHERE 1 <(SELECT COUNT (*) FROM incr WHERE empno = emp.empno);

CHAPTER 13

Agricultural Databases, Technical Reports and Conference Reports

Agriculture Database

Agriculture Database Concepts and Importance

The Necessity for the Construction of Agricultural Database

The construction of database is the basic work for the construction of the digital library and there must be a lot of databases that support the resources for the operation of the digital library. The necessity for the construction of the agricultural database is based on three aspects:

1. Diversified information requirements

Large-scale integrated databases have become the indispensable information sources for the teaching and research work in the agriculture colleges. Because of the different groups of readers, course fields, research direction and information utilization purposes, people have different requirement for information resources. As the variety of the social information demand, the integrated information database developed still cannot meet the demand for the readers' research work in some field, while the agricultural database may meet the demand in the agricultural field.

2. Competition in information industry

The rapid development of information network and automation leads to the development of information industry and consultation services. The development of the social information industry makes the service targets, contents; efficiency and existing space face the unprecedented challenge. The construction of various agricultural databases which the main service object demand, these are to keep its own characters, strength and expand the existing space under the internet and digital surroundings.

3. Benefits of information resources

Agricultural database may offer the important information guarantee to the development for the agricultural economic construction, the agricultural scientific research and the agricultural education. The agricultural database offers accurate, efficient and important information to the agricultural scientific research. It saves time of workers with science and research, avoid repeated work, and promote the science and research development. The agricultural database is also regarded as the foundation of large-sized and integrated database. The construction of the great deals of agricultural database with high quality may offer the good digital resources to the construction, updating and the development for the agricultural integrated database. The huge information, deep capacity of information service and the strong camp of information technology in the agriculture library, these decide that it is the authoritative position in the fields of the agricultural database.

The Orientation of Topic Selection for the Construction of the Agricultural Database

Under the surrounding of the Internet, the agriculture college's libraries should put the construction of its library information into the first place. We should know the latest teaching; the trends and requirements of scientific research in time and the collection of the digital information resources should be practical and scientific. The orientation of the selected topics for the agricultural database is the strategic problem and the following four major factors should be taken into consideration.

1. The selected topics should meet the requirement of specific groups of readers.

The resource construction, should take consideration into the requirement for its own specific readers. The library should set up the characterized database that can meet with the requirement for readers at most. In order to the requirements for the consumers the contents of the special database must integrate the most complete and the latest results for subjects or majors according to specific services object's specific demands, which can solve the difference between the unlimited information and the limited consumers' requirement.

2. The selected topics should meet the demand for the development of key specialized courses

In order to adapt the economic development and complete the mission for the agriculture talents, the agriculture colleges should choose some subjects. They take effective measures to construct the main subjects. Thus, they can reach the first level of the same subject at home and abroad on the teaching quality, scientific research and management. The selected topic for the agricultural database should give priority to the requirement for the important subjects, and has the consciousness of pioneering for the resources construction.

3. The selected topics should meet the demand for the science and research work

The agricultural database offers accurate, efficient and important information to the agriculture scientific research. It saves time of workers with science and research, avoid repeated work and promote the science and research development. Except setting the special topic as subjects, the special database also may cross the special topic and share the resources with others and promote the development of scientific study. The special database integrate huge information resources and is changed into the digital information before going to the internet, which improves the degree of information resources shares and meets with the demands for scientific research. In view of the advantages of special database, the library as the supplier and the manager, should take full consideration into the requirements of the scientific research work and offer literary information guarantees to the scientific research work.

4. The selected topics should meet the requirement for the agricultural production practice and agro-technology progress

The agriculture produce practice and technology progresses are very wide fields. The higher education should face the produce practice and promote technology progress and foster the qualified talents in this field. Thus, the resources constructions play an important role and have wide market, such as, the information database for agriculture education knowledge, for effecting production supply and marketing and for adjusting the production segment and essential factor digital and so on. This kind of database information resources has high accuracy and period of effectiveness.

The data collection comes from production practice or scientific experiment and suffers the inspection of production practice and market. Because this kind of the database information may have effect on the production and market segments, the college libraries should develop the special database information as soon as possible.

The Principle of Specialization in the Construction of the Agricultural Database

The specific database is to collect and arrange the library of some subject and form the production or achievement processed. Today, database is developing towards the direction of large scope and network. We should standardize the primary database design, the details and procedure, though self-building monographic database cannot reach the standard of the database made by the professional company. The quality of the proof, design, process, standard-making, data-collecting, is being controlled so as to ensure the quality of database and realize the share of network.

1. Having relevant key specialized courses for study or research

Colleges and universities have their own key subjects. Monographic database construction should reflect the characteristics of the key subjects and majors as well as consider the need of teach and study so as to serve the majors' construction and subjects' development.

2. Having readers of specific majors who need and use these resources

Students, teachers and science research workers are the main groups who use monographic database. Database's content should consider the need of target readers. The final purpose of constructing database is to benefit more readers. It makes no means without readers' demand.

3. Having the documentary base used for collecting specialized database resources

Database construction is a developing research work, but the content of database is a kind of integration for current resource. Documentary base used for collecting professional data is needed for database construction. This base includes resources and some other resources that conform to intellectual property right.

4. Having conditions for collection of relatively complete specialized information

Database construction is not only to classify and integrate data, but also to provide systematic knowledge and results for readers, so the specific research results or professional resources of the library or institute should be the main content of database. As a result, reader can get valuable materials from the database.

5. Having professional workers

In order to fit the professional demand, selected topic should reflect the advantages of the major and professional workers. Other workers of relative subject can also be invited to manufacture the database.

The Principle of Specificity in Database Construction

The continuity of database construction is as follows:

1. The specificity of collected materials

The collected information of the monographic database should conform to the major and the specific topic. It should be avoided that collecting non-related information for expanding the scale of the database.

2. The specificity of the database structure

The specificity of the database structure in the same style database can facilitate the makers and the reader as well as lower the cost. Database construction needs support of some software, but it costs much to develop a software or buy a software, so the database software should not be changed as chosen.

3. The specificity of making procedure

The specificity of making procedure can simplify working procedure so that the maker can grasp the skills of construction. Thus, working effectiveness is improved.

4. The specificity of the way of retrieval

The specificity of search ways can simplify the use ways, whereas readers cannot concentrate to get information from database, but spend lots of time on learning how to use database.

Top Quality Principle of Building Databases

People are dissatisfied in the face of large-scale colligating databases because of the numerous irrelevant information. As the complementarities and choiceness of the large-scale colligate commercial databases, self-built special databases are to satisfy the special study, special research and produce practice of special readers. While building them, we should follow the top quality principle that fining the materials, finding accurate data, and providing profound content.

1. Fining the materials can make special databases small-scaled but practicable

It is especially important to obey this principle while building literature databases for special study. Builders must familiarize or expert at relevant special knowledge, therefore, they can collect authoritative materials with high quality to make each record in the databases effective and useful.

2. The accuracy of data can improve the credibility and reliability of the special databases

This is especially important for building information or numeric databases with consult and direct meaning in produce and circulate area. The accuracy of data is decided by collecting occasion (ex. Market value information), collecting ways and means (ex. real time transfer on line), collecting method (ex. scientific data from exam) and the sense of duty of the people who collect data. The numeric database used to guide produce practice, such as the information to accommodate different produce step and essential factor, must suffer from produce practice.

3. Profound content can enhance the authority and guidance of the special databases

This point is much more important to databases for science researching. The content of the databases must be profound enough to reflect the leading field of the subject, hotspot and the latest achievement of the relevant research. Also it must satisfy some certain kinds of demands. We must add images, charts, audio, video and real time online information to provide information with multi-angle of view instead of single literal information.

Conclusion

An agricultural library should have the ability of information accumulation and information services, which is mostly embodied by special databases under network circumstances. However, the agricultural databases would have no advantages in information service without special agricultural databases. Special agricultural databases occupy a particular position and play an important role in the construction of database for agricultural libraries. They can make up the insufficiency of bibliography, digest database and integrated full text database. They can provide the information regarding the latest research achievements, research level and trends of relevant specialized courses, so that users can get conveniently first-hand materials so as to adjust research direction and create new research results.

We should carefully choose the content of special databases that can satisfy the requirement of readers or users and stress subject features, professional features and area features of special subjects. With dedicated and refined work, the agricultural database with a larger amount of information will attract more users, have its own position in the information resources of network and achieve good social and economic benefits.

Agriculture Database Types

AGRICOLA

This is a bibliographic database of citations to the agricultural literature created by the National Agricultural Library and its cooperators. Production of these records in electronic form began in 1970, but the database covers materials in all formats including printed works from the 15th century. The records describe publications and resources encompassing all aspects of agriculture and allied disciplines, including animal and veterinary sciences, entomology, plant sciences, forestry, aquaculture and fisheries, farming and farming systems, agricultural economics, extension and education, food and human nutrition, and earth and environmental sciences.

Agricultural Research Service Water Database

This indicates the collection of precipitation and stream flow data from small agricultural watersheds in the United States.

AQUASTAT

This agricultural database gives "The Food and Agriculture Organization's (FAO)" global information system of water and agriculture developed by the Land and Water Development Division of FAO. The objective of AQUASTAT is to provide users with comprehensive information on the state of agricultural water management across the world, with emphasis on developing countries and countries in transition.

Measured Annual Nutrient Loads from Agricultural Environments (MANAGE) Database

Compilation of "measured annual nitrogen (N) and phosphorus (P) load data representing field-scale transport from agricultural land uses." Includes "measured nutrient load data and corresponding watershed characteristics; information to support water quality management, modeling, and future research design; and a platform following user input of additional project-specific data."

National Wetlands Research Center Publications and Information Products

Database of studies and reports from this organization within the USGS. *Note*: The database may not display properly when using Netscape.

PAN Pesticide Database

"Brings together a diverse array of information on pesticides from many different sources, providing human toxicity (chronic and acute), eco-toxicity and regulatory information for about 6,400 pesticide active ingredients and their transformation products, as well as adjuvants and solvents used in pesticide products."

Pesticide Properties Database

A compendium of chemical and physical properties of 334 widely used pesticides. Information included in the database focuses on 16 of the most important properties that affect pesticide transport and degradation characteristics.

Plant Science

A bibliographic database containing citations and abstracts of scientific literature on plant science, focusing on all plant scientific aspects, especially on pathology, symbiosis, biochemistry, genetics, biotechnology, techniques and environmental biology. From July 1996 onwards, informative abstracts have been added when available. Provides links to web sites on related topics.

Safe Drinking Water Information System

The safe drinking water information system provides access to information about Safe Drinking Water Act violations and enforcement history for public water supplies during the last ten years.

Agriculture Libraries

AGRICOLA (AGRICULTURAL ONLINE ACCESS) is a bibiliographic database to the agricultural literature created by the national Agriculture library (NAL). Resources for all aspects of agriculture and allied disciplines including animal and veterinary science, entomology, farming & farming systems, agricultural economics, extension & education, earth and environmental science.

AGRICOLA database is organized into two data sets.

1. NAL online public Access Catalog, known as "NAL Catalog" contains books, audiovisual materials, serial publications and other materials in the NAL collection.
2. "Journal Articles" includes journal article and reports

Website: http://www.nal.usda.gov/ag98/ag98.html.

International Rice Research Institute (IRRI) Library

IRRI's library is the world's leading rice library with more than 1,00,000 book entries about 2000 journals.

Website: http://ricelab.irri.cgiar.org/

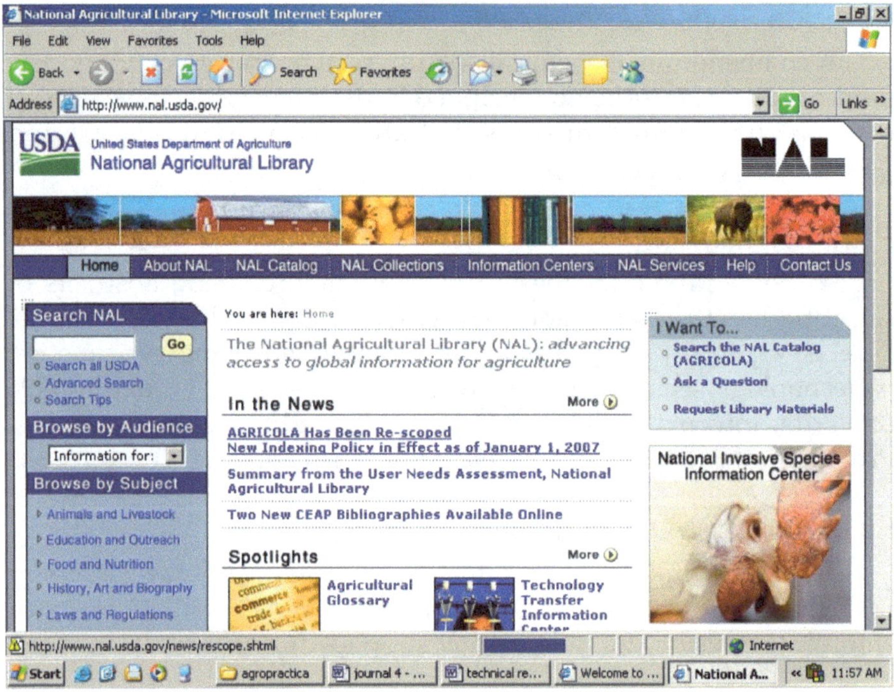

Agricultural Market Information Virtual Library

This website serves as a guide to the different sources as market information presently available as the Internet. These sources may be daily, weekly, monthly or yearly reports, outlook reports, commodity reports or quotes from different Market information sources Available through the internet are daily to yearly market report, outlook reports, prices, commodities and quotes.

Website: http://www.aec.msu.edu/agecon/fs2/market/content.html.

Agrifor

Website: http://agrifor.ac.uk/

Agrifor is a gateway to evaluated , quality internet resources in agriculture, food and forestry, aimed students, researchers, academics and practitioners in agriculture, food or forestry.

Aginfonet aglibrary

Aglibrary takes a new approach to the delivery of electronic agricultural information.

Website: http://www.aginfonet.com

Journals

Journals include

- Scholarly literature
- Peer-reviewed papers
- Thesis
- Books
- Preprints
- Abstracts and
- Technical reports from all broad areas of research.

E-Journal

An e-journal is an article or complete journal available electronically via a website on the internet. It could be available free or as a part of a paid service.

Agriculture Journals

Books and journal publishing (both print & electronic) includes preparation of a number of practical manuals and hand books and facilitation of workshops.

Journal Online Programmes

The journal online Programmes aim to enable the research undertaken and published in developing and transitional countries to become member widely known, accessible and to strengthen the academic and scholarly publishing sectors. For example, African Journals Online (AJOL)

Website: http://www.inasp.info/ajol/index.html.

Lists of Journals available:

- Agricultural Science and Resources Management
- African Crop Science Journal
- African Journal of Livestock Extension.
- Agro-Science.
- Journal of Agriculture and Social Research.
- Tropical Veterinarian.

- The Water Wheel.
- International Journal of Agriculture & Rural Development
- Journal of Environmental Extension.
- Global Journal of Agricultural Science.
- Global Journal of Environmental Extension.

The Journal of Agro Biotechnology, Management & Economics:

Website: http://www.tamu-commerce.edu/library/libsci/fulltext/jounals-ag.html/

The following websites are available.

- AgExporter
- Agricultural and forest meteorology.
- Agricultural Economics.
- Agricultural System
- Agricultural water management
- Animal Reproduction Science
- Biological Control.

Computers and Electronics in Agriculture

Website: http://www.science direct.com/science/journal/01681699.

Science direct in publishing the journal "Computers and Electronics in Agriculture".

Times Agricultural Journal

Website: http://www.etagriculture.com

Times Agriculture journal is targeted at decision makes in companies involved in food and agro-chemical business, non-government organizations, departments of agriculture, horticulture, floriculture etc.

Agriculture 21

Agriculture 12 is the online magazine published by the Agriculture department of FAO (Food and Agriculture Organization)

Website: http://www.fao.org/ag

The Virtual Academy for Semiarid Tropics (VUSAT)

Website: http://www.vusat.org

VUSAT aims to mobilize communities and intermediaries of the dry tropics by sharing information's, knowledge and skills related to climate literacy, drought preparedness and best practices in dry land agriculture and other relevant issues.

Rice Web

The rice web provides details about Rice on the following topics

- History and general information
- Feeding the world
- Production and trade
- Environment of rice farms
- The plant and how it grows
- Geography of production
- Research
- World rice statistics
- Terminology and literature

Website: http://www.riceworld.org/contents.htm

Rice World

Website: http://www.riceworld.org/

Rice world is located at the International Rice Research Institute (IRRI)

This provides the details about

- Insects with microscope viewing
- Farmers clothing
- Traditional rice farming implements
- Rice products
- Wild rice
- Display of the four rice ecosystem
- High yielding rice varieties

IRRI Knowledge Bank

Website: http://www.knowledgebank.irri.org

The Rice knowledge Databank provides tools for the benefit of growers and researcher, they are in Hindi and Tamil.

- *Rice Doctor*: A tool for diagnosing your field problem.
- *TROP RICE*: A tool for making crop management decision.

CABI publication

Website: http:// www.cabi-publising.org/plantscience.asp

CABI publishing delivers high-quality information an all aspects of the plant sciences, including biotechnology, crop protection, forestry, horticulture, mycology and plant biology.

TECHNICAL REPORTS

What is a Technical Report?

A **technical report** (also: **scientific report**) is a document that describes the process, progress, or results of technical or scientific research, or the state of a technical or scientific research problem. Unlike other scientific literature, such as scientific journals and the proceedings of some academic conferences, technical reports rarely undergo comprehensive independent peer review before publication. Where there is a review process, it is often limited to within the originating organization. Similarly, there is no formal publishing procedure for such reports, except where established locally.

General characteristic of Technical Reports

- The technical reports contain:
- contain Graphics
- contain very detailed and factual data
- document information sources
- Defined for a realistic group of readers

Online gardening information

http://www.gardenguides.com/

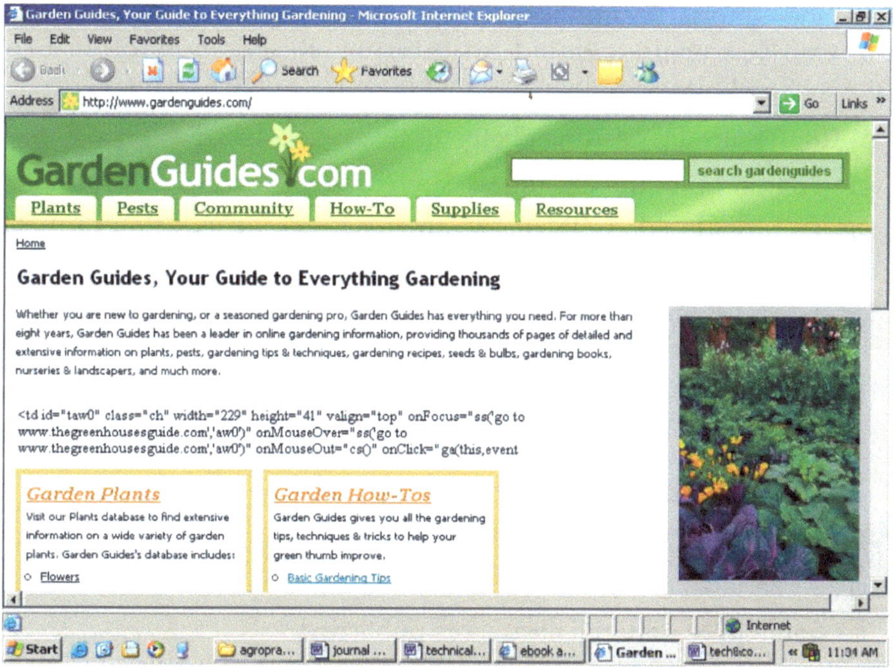

Water resource institute

http://twri.tamu.edu/reports.php

Wetland

http://www.epa.gov/owow/wetlands/guidance/

Resource Management Technical Reports are being made available electronically by the Department of Agriculture and Food Western Australia.

http://agspsrv34.agric.wa.gov.au/environment/publications/RMtechreports/index.htm

CONFERENCE PAPERS

What are Conference Papers?

Conference papers are the collection of academic papers that are published in the context of an academic conference. They are usually distributed as printed books at the end of the conference. This printed book is called the conference proceedings. Proceedings contain the contributions made by the researchers at the conference. They are written record of the work that is presented to fellow researchers.

The collection of papers is organized by one or more persons, who form the editorial team. Proceedings are also published in electronic form on CD or distributed on Internet. Researchers often use conferences to present their research findings to their peers.

Conference Proceedings helps us:

- To find up to the minute research on a topic
- To provide you with the researchers' own explanation of their research
- To find sources of information related to this research e.g. subject experts, research team members
- To find out about best practices or innovations in a particular area.

Where would we find them?

- As separate, one-time publications
- As serials
- Sometimes only the abstract is published
- Sometimes the paper may be obtained by only contacting the author or the society.
- Sometimes the full text is available on the web.
- Many individuals publish their conference papers on their own personal websites, or on the websites of their research institution.
- Libraries are also holding some hard copy of conference proceedings or indexes (searchable lists) of conference proceedings.

- Some electronic databases include conference papers.
- Journals often produce individual issues containing entire proceedings.
- Conference proceedings may be published individually as a monograph.
- Occasionally the papers are never published but may be cited by authors who attended the meetings. As a result of these variations, locating conference proceedings can be complicated.

Knowledge Base on Water, Food and Environment

http://www.iwmi.cgiar.org/assessment/Publications/conference_papers.htm

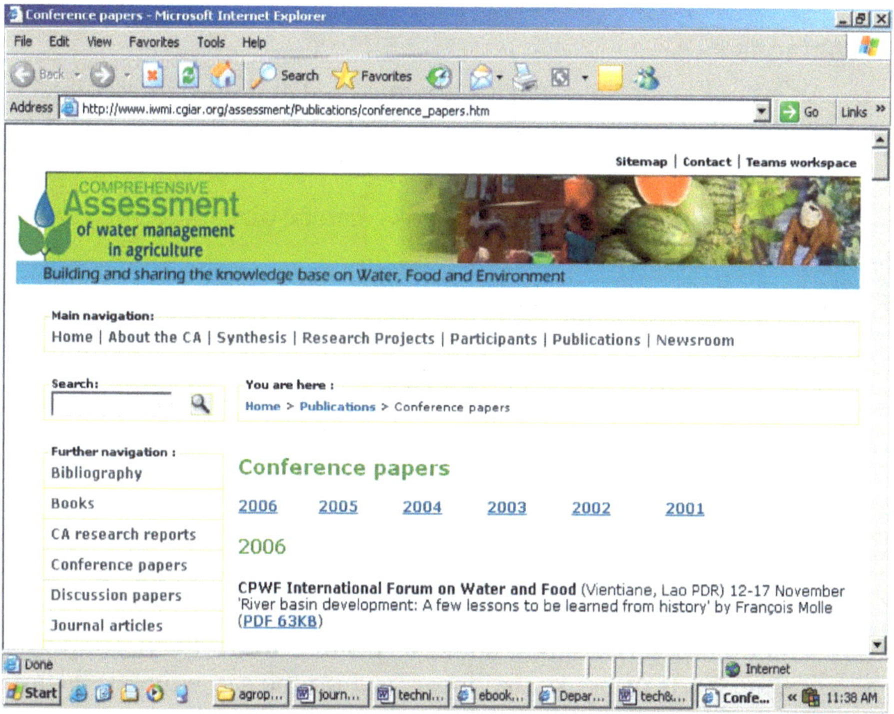

Agriculture Economics

http://www.aec.msu.edu/fs2/kenya/c_papers.htm

Agricultural Information

http://www.fao.org/aims/pub_conf.jsp

E-BOOK

An eBook is an electronic version of a traditional print book that can be read by using a personal computer. An e-book (short for electronic book, also written eBook or ebook), also known as a digital book, is an e-text that forms the digital media equivalent of a conventional printed book, sometimes restricted with a digital rights management system. An E-book, as defined by the Oxford Dictionary of English, is "an electronic version of a printed book which can be read on a personal computer or hand-held device designed specifically for this purpose. E-books are usually read on dedicated hardware devices known as e-Readers or e-book devices. Some personal computers and cell phones can also be used, especially to read documents in pdf format.

Advantages

- An e-book can be purchased, downloaded and used immediately, whereas when one buys a book one has to go to a bookshop or wait for a delivery.
- Educational text books (e-textbooks) can be distributed at a lower cost than print versions per student.
- Second generation e-books (and color e-readers) have motion capability.
- E-Books can be printed for less than the price of traditional new books using new on demand book printers.
- Text can be searched automatically and cross-referenced using hyperlinks.
- A single e-reader containing several books is easier to carry around (less weight and volume) than the same books (or sometimes even a single book) in printed form. Even hundreds or thousands of books may be stored on the same device. Using removable media even more can be carried around easily.
- Also at a fixed place such as at home it can be an advantage that an e-book collection requires very little space.
- Mobile availability of e-books may be provided for users with a mobile data connection, so that these e-books need not be carried around.

- E-books can allow non-permanent highlighting and annotation.
- E-book websites can include the ability to translate books into many different languages.
- Many E-book readers have the ability to enlarge fonts past standard large-print size.
- E-books may allow animated images or multimedia clips to be embedded.
- E-books allow for greater fidelity in color reproduction compared to CMYK color printing (although most e-book readers have only monochrome displays).
- Depending on the device an e-book may be readable in low light or even total darkness. For devices for which this applies, energy consumption for reading without daylight is less than that of a lamp needed for reading a printed book.

An e-book can automatically open at the last read page.

- While an e-book reader costs much more than one book, the electronic texts are generally cheaper. Moreover, a great share of books are available free of charge.
- Text-to-speech software can be used to read the text. However, the voice will be auto-generated, thus the quality worse than audio books.
- An e-book can be offered indefinitely, without ever going "out of print".
- Depending on possible digital rights management, it may be easy and cheap to produce a back-up for the case that the e-book is lost or damaged, and/or it may be possible to get a free new copy if that happens.
- It is easier for authors to self-publish e-books.
- A free e-book can stimulate the sales of the printed version.
- The production of e-books does not consume paper, ink, etc. Printed books use 3 times more raw materials and 78 times more water to produce.

Drawbacks

- Reading e-books requires newer and upgraded electronic devices and changing software. Even in the case of reading it on a personal computer one already has, it may require additional software.
- A paper book will never turn off, unlike an eBook.
- Only a small fraction of all books are available as eBooks. Most books will not be eBooks for some time.
- Most eBooks will not have all the graphics, photos, illustrated covers, quality paper and aesthetic appreciation of binding.
- Paper books can be bought and wrapped for a present, unlike an eBook.
- A library of books is visually appealing, unlike the data storage of eBooks.
- A small paperback is easier to carry around than a typical e-book reader.
- E-book readers require electrical power; in the case of mobile use, the battery will eventually get exhausted.
- A book will never turn off or be unusable. A book is long lasting, especially in a remote area without electricity.
- E-book readers are more susceptible to damage from being dropped or hit.

E-book readers can malfunction and e-books can be damaged due to faults in hardware or software.

- E-books may cause people "to do the grazing and quick reading that screens enable, rather than be by themselves with the author's ideas."
- People may use E-books simply for reference purposes rather than for extended reading for pleasure and leisure.
- E-book readers are more likely to be stolen than individual paper books.
- A few used books are far more affordable. Hundreds of used books can be purchased for the same price as an e-book reader.

- Depending on the device, an e-book may be difficult to read in bright sunlight.
- Most publishers don't produce the e-book equivalent of their printed books. In other cases the product quality is lower or it is released later.
- E-books can be easily hacked through the use of hardware or software modifications and widely disseminated on the Internet and/or other e-book readers, without approval from the author or publisher. This ease of piracy is a significant drawback for publishers.
- If an e-book device is stolen, lost or broken beyond repair, all e-books stored on the device may be lost (Although this may be avoided by backups either on another device or by the e-book provider).
- There is a loss of tactility and aesthetics of book-bindings. Also lost is the ability to very quickly riffle through the pages to search for a particular section or to get a sense of the book merely by sight.
- Screen resolution of reading devices is currently lower than actual paper.
- Due to the digital rights management, customers cannot resell or loan their e-books to other readers.
- Some books available as e-book cannot be read on some e-book readers because they are not supplied in a format those readers allow.
- While printed books remain readable for ages, changing technologies and less durable electronic storage media require e-books to be copied to a new carrier after some years.
- E-book readers require various toxic substances to produce, and are an environmental hazard as they are non-biodegradable.

Physical discomfort for some users, including eye strain.

- One will never read thousands of books really well or over a short period of time, so the high amount held on an eBook reader becomes irrelevant (Though it might be useful if the limited amount of eBooks is used as a partial reference library).

- Some E-book formats cannot be easily backed up. If a book is accidentally lost or deleted, we must repurchase the book.
- A book is relatively safe and durable from electromagnetic pulses, overloads, water, strong winds, falls, and accidents.
- Books are more readily available for poor people to read. Expensive technology is not readily available to most of humanity.
- A book on paper cannot be destroyed easily or damaged by a sharp blow.
- An e-book reader needs to be protected against extreme cold, heat, water, sand, and weather.
- A small percentage of readers will use e-books, as paper books still remain the most popular choice so far.

E-Books in Agriculture

1. Agronomy and Crop sceince

http://www.diesel-ebooks.com/cgi-bin/category/TEC003000/Agriculture.html

2. Gardening e-book

http://www.earthlypursuits.com/AllotGuide/AllotGuide.htm

3. Agriculture e-book

http://www.ebooks.com/subjects/b/539/1.asp

Exercise to Practice

I. Visit this entire website Journals

1. *The journal of Agro biotechnology, management & Economics*

   Website : www.agbioforum.org

2. *Computers and Electronics in Agriculture*

   Website: http://www.sciencedirect.com/science/journal/01681699

3. *Times Agricultural Journal*

   Website: http://www.etagriculture.com

4. *Agriculture 21*

   Website: http://www.fao.org/ag

II. Visit this website Libraries

1. *International Rice Research Institute (IRRI) Library*

   Website: http://ricelib.irri.cgiar.org/

2. *Agricultural Market Information Virtual Library*

   Website:http://www.aec.msu.edu/agecon/fs2/market/content.html.

3. *Agrifor*

   Website: http://agrifor.ac.uk/

4. *Aginfonet aglibrary*

   Website: http://www.aginfonet.com

III. Visit this entire website to view Technical reports

1. *Water resource institute*

   http://twri.tamu.edu/reports.php

2. *Wetland*

   http://www.epa.gov/owow/wetlands/guidance/

3. *Resource Management Technical Reports are being made available electronically by the Department of Agriculture and Food Western Australia.*

   http://agspsrv34.agric.wa.gov.au/environment/publications/RMtechreports/index.htm

IV. Visit this entire website to get conference paper

1. *Agriculture Economics*

   http://www.aec.msu.edu/fs2/kenya/c_papers.htm

2. *Agricultural Information*

   http://www.fao.org/aims/pub_conf.jsp

3. *Rice World*

   http://www.riceworld.org

4. *IRRI Knowledge Bank*
   http://www.knowledgebank.irri.org

5. *CABI publication*
   http:// www.cabi-publishing.org/plantscience.asp

V. Explain the importance of agricultural database.
   List the entire agricultural database.

CHAPTER 14

Web Portal

Web Portal

A web portal, also known as a links page, presents information from diverse sources in a unified way. Apart from the standard search engine feature, web portals offer other services such as e-mail, news, stock prices, information, databases and entertainment. Portals provide a way for enterprises to provide a consistent look and feel with access control and procedures for multiple applications and databases, which otherwise would have been different entities altogether. Examples of public web portals are MSNBC, Yahoo!, AOL, iGoogle and Netvibes

History

In the late 1990's the web portal was a hot commodity. After the proliferation of web browsers in the late-1990s many companies tried to build or acquire a portal, to have a piece of the Internet market. The web portal gained special attention because it was, for many users, the starting point of their web browser. Netscape became a part of America Online, the Walt Disney Company launched Go.com, and Excite and @Home became a part of AT&T during the late 1990s.

Lycos was said to be a good target for other media companies such as CBS.

Types of Portals

Horizontal vs. Vertical Portal

Two broad categorizations of portals are horizontal portals, which cover many areas, and vertical portals, which are focused on one functional area. Another definition for a horizontal portal is that it is used as a platform to several companies in the same economic sector or to the same type of manufacturers or distributors. A vertical portal consequently is a specialized entry point to a specific market or industry niche, subject area, or interest, also called vortal.

Personal Portals

A personal portal is a site on the World Wide Web that typically provides personalized capabilities to its visitors, providing a pathway to other content. It is designed to use distributed applications, different numbers and types of middleware and hardware to provide services from a number of different sources. In addition, business portals are designed to share collaboration in workplaces. A further business-driven requirement of portals is that the content be able to work on multiple platforms such as personal computers, personal digital assistants (PDAs), and cell phones/mobile phones. Information, news, and updates are examples of content that would be delivered through such a portal.

Personal portals can be related to any specific topic such as providing friend information on a social network or providing links to outside content that may help others beyond your reach of services. Portals are not limited to simply providing links. Information or content that you are putting on the internet creates a portal, or a path to new knowledge and/or capabilities.

Regional Web Portals

Along with the development and success of international personal portals such as Yahoo, regional variants have also sprung up. Some regional portals contain local information such as weather forecasts, street maps and local business information. Another notable expansion over the past couple of years is the move into formerly unthinkable markets.

"Local content - global reach" portals have emerged not only from countries like Korea (Naver), India (Rediff), China (Sina.com), Romania, Greece (in.gr) and Italy, but in countries like Vietnam where they are very important for learning how to apply e-commerce, e-government, etc. Such portals reach out to the widespread across the world.

Government Web Portals

At the end of the dot-com boom in the 1990s, many governments had already committed to creating portal sites for their citizens. In the United States the main portal is USA.gov in English and GobiernoUSA.gov in Spanish in addition to portals developed for specific audiences such as Disability.gov; in the United Kingdom the main portals are Directgov (for citizens) and businesslink.gov.uk (for businesses).

The official web portal of the European Union is Europa (web portal). Europa links to all EU agencies and institutions in addition to press releases and audiovisual content from press conferences. All relevant health topics from across Europe are gathered in the Health-EU portal.

Corporate Web Portals

Corporate intranets became common during the 1990s. As intranets grew in size and complexity, webmasters were faced with increasing content and user management challenges. A consolidated view of company information was judged insufficient; users wanted personalization and customization. Webmasters, if skilled enough, were able to offer some capabilities, but for the most part ended up driving users away from using the intranet.

Many companies began to offer tools to help webmasters manage their data, applications and information more easily, and through personalized views. Portal solutions can also include workflow management, collaboration between work groups, and policy-managed content publication. Most can allow internal and external access to specific corporate information using secure authentication or single sign-on.

Java Specification Request (JSR) 168 standards allow the interoperability of portlets across different portal platforms. These standards allow portal developers, administrators and consumers to integrate standards-based portals and portlets across a variety of vendor solutions.

The concept of content aggregation seems to still gain momentum and portal solution will likely continue to evolve significantly over the next few years.

With the increase in user generated content, disparate data silos, and file formats, information architects and taxonomist will be required to allow users the ability to tag (classify) the data. This will ultimately cause a ripple effect where users will also be generating ad hoc navigation and information flows.

Corporate Portals also offer customers & employees self-service opportunities.

Hosted Web Portals

As corporate portals gained popularity a number of companies began offering them as a hosted service. The hosted portal market fundamentally changed the composition of portals.

In many ways they served simply as a tool for publishing information instead of the loftier goals of integrating legacy applications or presenting correlated data from distributed databases. The early hosted portal companies such as Hyperoffice.com or the now defunct InternetPortal.com focused on collaboration and scheduling in addition to the distribution of corporate data.

As hosted web portals have risen in popularity their feature set has grown to include hosted databases, document management, email, discussion forums and more. Hosted portals automatically personalize the content generated from their modules to provide a personalized experience to their users. In this regard they have remained true to the original goals of the earlier corporate web portals.

Domain-specific Portals

A number of portals have come about that are specific to the particular domain, offering access to related companies and services, a prime example of this trend would be the growth in property portals that give access to services such as estate agents and removal firm. Along the same lines, industry-specific news and information portals have appeared, such as the clinical trials specific portal: IFPMA Clinical Trials Portal.

Exercise to Practice

1. Define web portal.
2. Explain corporate web portal with an example.
3. What are the types of web portals?
4. Give an example for government web portals.

CHAPTER 15

Introduction to Bioinformatics

Definition of Bioinformatics

- Bioinformatics is the computational branch of molecular biology.
- The science of developing computer software and algorithms to record and analyze biology related data.
- Bioinformatics is the application of computer technology to the management of biological information.

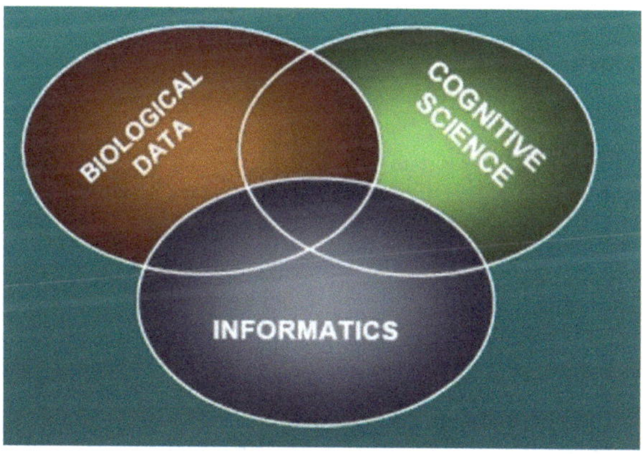

"The collection, archiving, organization and interpretation of biological data."

Bioinformatics deals with :

- Large biological data sets
- Computer representations and algorithms
- Storage, analysis, visualization, prediction, and design

History of Bioinformatics

The bioinformatics has come into existence due to growth in molecular biology, technology and computers in the last three decades. Alfred Sanger earned his first Nobel Prize for sequencing of insulin inaugurated the modern era of molecular and structural biology.

Genesis of Bioinformatics

In the pre-computer era (1960s) known protein sequences accumulated were manipulated manually. The new area of research

- The analysis of protein sequences using computers.
- The genesis of bioinformatics.

Structural Bioinformatics

First 3D structure of a protein was developed in the year 1958 by Drs. Kendrew and Perutz.

Structural bioinformatics is the branch of *bioinformatics* which is related to the analysis and prediction of the three-dimensional structure of biological *macromolecules* such as *proteins, RNA,* and *DNA*. It deals with generalizations about macromolecular 3D structure such as comparisons of overall folds and local motifs, principles of molecular folding, evolution, and binding interactions, and structure/function relationships, working both from experimentally solved structures and from computational models.

Bioinformatics field devoted the development of tools to navigate between sequences and 3-D structures called structural bioinformatics. The first truly efficient technique to sequence DNA was discovered in 1977. The term structural has the same meaning as in *structural biology,* and structural bioinformatics can be seen as *computational* structural biology.

Genomics

Genomics: describe the scientific discipline of mapping, sequencing and analysing genes. In 1995, the first sequence of an entire genome was determined.

New Branch of Bioinformatics

Parsing of large DNA sequences into their components

- Genes
- Transcription units
- Protein coding regions
- Regulatory elements and so forth
- Genome annotation

Use of Bioinformatics in Agriculture-Industries like Pharmaceuticals and Business

Bioinformatics is being used in following fields:

- Molecular medicine
- Personalised medicine
- Preventative medicine
- Gene therapy
- Drug development
- Microbial genome applications
- Waste cleanup
- Climate change Studies
- Alternative energy sources
- Biotechnology
- Antibiotic resistance
- Forensic analysis of microbes
- Bio-weapon creation
- Evolutionary studies
- Crop improvement
- Insect resistance
- Improve nutritional quality
- Development of Drought resistance varieties
- Veterinary Science

Bioinformatics - Application in Agriculture

Sequence the genomes of plants. Search for the genes to study the functions through bioinformatics tools to produce stronger, more

- Drought,
- Disease and
- Insect resistant crops.

Crop Improvement

Comparative genetics of the plant genomes has shown that the organization of their genes has remained more conserved over evolutionary time than was previously believed. These findings suggest that information obtained from the model crop systems can be used to suggest improvements to other food crops.

Insect Resistance

Genes from Bacillus thuringiensis that can control a number of serious pests have been successfully transferred to cotton, maize and potatoes. This new ability of the plants to resist insect attack means that the amount of insecticides being used can be reduced and hence the nutritional quality of the crops is increased.

Improve Nutritional Quality

Scientists have recently succeeded in transferring genes into rice to increase levels of Vitamin A, iron and other micronutrients. This work could have a profound impact in reducing occurrences of blindness and anaemia caused by deficiencies in Vitamin A and iron respectively. Scientists have inserted a gene from yeast into the tomato, and the result is a plant whose fruit stays longer on the vine and has an extended shelf life.

Grow in Poorer Soils and Drought Resistant

Progress has been made in developing cereal varieties that have a greater tolerance for soil alkalinity, free aluminium and iron toxicities. These varieties will allow agriculture to succeed in poorer soil areas, thus adding more land to the global production base. Research is also in progress to produce crop varieties capable of tolerating reduced water conditions.

Veterinary Science

Sequencing projects of many farm animals including cows, pigs and sheep are now well under way in the hope that a better understanding of the biology of these organisms will have huge impacts for improving the production and health of livestock and ultimately have benefits for human nutrition.

Drug Development

The pharmaceutical industry has embraced genomics as a source of drug targets. It also recognizes that the field of bioinformatics is crucial for validating these potential drug targets and for determining which ones are the most suitable for entering the drug development pipeline.

In the past, new synthetic organic molecules were tested in animals or in whole organ preparations. This has been replaced with a molecular target approach in which in-vitro screening of compounds against purified, recombinant proteins or genetically modified cell lines are carried out with a high throughput. This change has come about as a consequence of better and ever improving knowledge of the molecular basis of disease.

For example, when we want to develop a broad-spectrum antibiotic, we are looking for targets that are present in a large number of bacteria yet have no similar homologues in human. This means that the antibiotic will be effective against many bacteria killing them while causing no harm to the human. In order to determine the role our potential drug target plays in a particular disease mechanism, we use DNA and protein chips. These chips can measure the amount of transcript or protein expressed by a cell at different times or in different states (healthy versus diseased). Clustering algorithms are used to organize this expression data into different biologically relevant clusters. We can then compare the expression profiles from the diseased and healthy cells to help us understand the role our gene or protein plays in a disease process.

Drug Designing includes:

- Clinical Trial

  Discovering a chemical compound and bringing it into a drug is a hard and time consuming process.

Two methods were followed for discovering the drugs namely

- Empirical method
- Rational method

Empirical Method

The empirical method is a blind and hit or loses method. Thousands of chemical compounds are tested on the disease without even knowing. The target on which the drug acts and Target is the disease cause and the mechanism of action is the discovery of penicillin made through empirical method.

Rational Approach

The advent of bioinformatics changed the empirical to rational. This approach starts from the clear knowledge of the target and the mechanism by which it is attacked.

Therefore drug discovery involves

1. Target research

   Target is the disease cause
2. Lead research

   Lead is the drug

Use of bioinformatics in pharmaceutical business :

Pharmaceutical industry recognizes the field of bioinformatics for

- Validating the potential drug targets and
- For drug (lead) discovery.

Target Example

- Acanthamoeba karatitis is an infective blinding eye disease, caused by a protozoan *Acanthamoeba.*
- Contact lenses or bathing in dirty swimming pools are the sources of infection.
- The amobea multiply on the cornea of the eye and invade deeply.
- Amoebae are eukaryotes. (a cell or organism with a distinct membrane-bound nucleus..)
- Any drug that can kill amoebae must invariably have side effects on human beings.

- Surprisingly protozoans carry plastids in their cytoplasm which are normally found in plants.
- Amoebae depend on these plastids for crucial fat metabolism, without which they cannot survive.
- The lipid metabolism of plastids is distinct from human lipid metabolism.
- Therefore the lipid metabolism in the plastids of amoebae is the target.

Target Discovery

- Immaterial of the biological system, the targets are usually the biomolecules, such as Enzymes, Receptors or ion channels. In the case of Acanthamoeba keratitis the target may be an enzyme.
- The validity of the enzyme depends on how much it is important for the survival of the pathogen.
- If the target is less significant it has no value.
- If the target is located inside the human system, the fluctuation of the target activity must correspond to the fluctuation of the disease severity.
- Under the study of bioinformatics field a sufficient level of confidence on the validity of the target is established.
- Once the target is identified a lead compound that can modulate the target can be developed.
- The database or software for chemical compounds are screened for the desirable molecule.
- The lead compound is built piece by piece using computer software.

  Eg: Hex

Analyzing Protein Sequence

Proteins are made up to basic building blocks called amino acid. Amino acids are complex organic molecules made of carbon, hydrogen, oxygen, nitrogen and sulfur atoms. Twenty amino acid are Alamine, glycine, tyrogine, glutamine etc.,

Amino acids are linked together as a chain. The first amino acid sequence of a protein is insulin 1951 with 110 resides.

Insulin = MALWMRLLPLLALLALWGPDPAAAFVN
QHLCGSHLVEALYLVLGERGFFYTPKJRREAE
DLQVGQVELGGGPGAGSLQPLALEGSLQKRGIVE
QCCTSICSLYQLENYCN

Need for Analysis

Most sensitive comparison are made at the protein level. Proteins are function abstracts of the genetic events that occur in DNA. Study of DNA Sequences underlines the detection of Phylogenetic set of sequence for same gene in individual of different species relationship through intron/Exon prediction, inference protein coding sequence, determination sequence through Open Reading Frame (ORF).

Analysis of DNA Sequence

A DNA Sequence or genetic sequence is a succession of letters representing the primary structure of real or hypothetical DNA molecule or stand with capacity to carry information.

The possible letters are A, C, G and T representing four nucleotide subunits of DNA Stand - adenine, cytosine, guanine, thymine bases covalently linked to phosphorous backbone. A succession of any number of nucleotides greater than four is liable to be called sequence. Sequence can be derived from the biological raw material through a process called DNA sequencing. DNA Sequence 4 Nucleotide compared to 20 amino acids.

Analyzing RNA Sequence

RNA differs from DNA by one nucleotides and RNA came a single strand. In DNA we has Thymine in RNA we have Uracil. U Instead of T. RNA sequence is encoded in DNA. So we work in RNA gene (DNA) Sequence rather than RNA Sequence.

RNA Structure

RNA molecules consist of a single strand nucleotides. Transporting and transmitting genetic information.

DNA into Protein

DNA sequence, you can translate it into the corresponding protein sequence by using genetic code. Four nucleotide to 20 amino acid.

From the given starting pointing your DNA sequence start reading the sequence 3 nucleotides cane triplet at a time current triplet is codon.

1. Read the DNA sequence ATGGAA GTATTTAAA GC GCCA CCTATTTAG.
2. Decompose it into successive triplet ATG GAA GTA.
3. Translate each triplet into the corresponding amino acid.

MEVFKAPPIGI STOP

ATG → Starting Codon.

TAA
TAG
TGA } Ending Codon.

Universal Genetic Code

Reading Frame: A sequence of codor beginning with an initiation codon and ending with termination codon with atleast 150 amino acid.

DNA in both the strand has six reading frame. Only 6 frame is used for any DNA coding region.

Open reading Frame: Largest reading frame uninterrupted by the stop codon (TAA TAG TGA).

Pairwise Alignment

Pairwise sequence alignment methods are used to find the best - Matching piecewise (local) or global alignments of two query sequences. Pairwise alignments can only be used between two sequences at a time, but they are efficient to calculate and are often used for methods a not requires extreme precision.

Multiple Sequence Alignment (MSA) is an extension of pairwise alignment to incorporate more than two sequences at a time. Multiple alignment methods try to align all of the sequences in a given query set.

Multiple Alignment

Arrangement of similar sequence is below the other so that the residues matches their position is called multiple alignment. A minimum of two sequence makes an alignment. But less than three sequences dar not give meaningful results.

V I I S C T G S S S N I G A G - N H V J W Y Q Q L P G

V T I S C T G S S N I G S --- I T V N W Y Q Q L P G

P E V T C V V D F Y P S G T A - E W E S N G - - - - -

Multiple alignments of sequence are very helpful for protein structure predictions. Multiple alignment of protein or DNA sequence reveals the hidden structural and functional features of unknown protein if the query protein sequence is aligned to a multiply aligned similar sequence.

Goal of MSA

It is to generate a concise information rich summary of sequence data in order to inform decision making or the relatedness of sequences to a gene family.

MSA – A Definition

MSA is a 20 table, in column the row represent individual sequence row the column the residue position. Sequence is laid into this grid in such a manner that,

a) The relative positioning of residues column in any one sequence is preserved.

b) Third residues in all the sequence are brought into vertical register.

1. Call the residue position in unaligned sequence - The absolute position.
2. By contract we call the aligned residue position the relative position.
3. All residues in any single column of an alignment will have the same position & almost certainly different absolute.
4. So, absolute position is a property of the sequences, while relative positions a property of alignment.

Multiple Alignments

- It is useful for predicting protein structure, also for predicting the functions of proteins is important to phylogenetic analysis.
- Better our multiple sequence alignment
- Required for better structural models
- Required for better functional predictions.

- Required for better phylogenetic trees outputs your sequences in their original order
  1. Click Run button at the button of the page
  2. Save your results

Results come in Three Sections like

1. Pairwise scores
2. Multiple alignments
3. The guide tree (not a phylogenetic tree)

Interpreting the MSA

(*) a star indicates an entirely conserved column

(:) a colon indicates columns where all the residues have roughly the same size

(.) a period indicates columns size for eg., our average good block is a unit atleast 10-30 a aminoacids long (*), five to seven colons (:) and a few periods sprinkled here and there.

Choosing the Right MSA Method

Culstal W

It uses a progressive method to build and alignment. Instead of aligning all the sequences at the same time it adds them one by one.

Running the EBI Clustal W Server

1. Point your browser to EBI clustalW server page at www. ebi.ac.uk/clustalw/index.html.
2. Paste the sequences you collected in sequences window
3. Choose fast from the alignment pull down menu
4. Use the output format pull-down menu to set selection of your choice.
   a. It is safe to use alignment without Numbers the default clustalW format.
5. Choose the input from the output pull - down when you set output order to input, clustal W.

Choose the Right Number of the Sequences

1. There's no absolute answer to this question
2. We should start with a relatively small no. of sequence - between 10 to 15 sequences is suitable for most cases.
3. After you get something interesting with this small set you can always increase its size
4. In any case, it hard to see any reason for generating a multiple alignment with > than 50 sequences.

If we Start 100s of Sequences, the Immediate Hit Troubles

1. Computing big alignment is difficult to public servers does not have infinite resources.
2. Building big alignment in difficult -multiple alignment programs are not very good at handling very large sets of sequences.
3. Displaying big alignment is difficult - you cant's print them and they clog your computer when you want to visualize them, if columns are longer than one page interpretation becomes impossible.
4. Using big alignment is difficult - free building and structure prediction programs cannot handle them easily.
5. Making accurate big alignment is difficult you want your multiple sequence alignment.

To be highly reliable so that you can be confident that all sequence it contains are true members of the family.

Main criteria for building a Multiple Sequence Alignment includes:

1. Structural Similarity

Amino Acids that play the same role in each structural are in the same column. Structure super position programs are only ones that use this criterion.

2. Evolutionary Similarity

Amino acid or nucleotides related to the same amino acid or nucleotide in the common ancestor of all the sequences are put in the same column. No automatic program explicity uses this criterion.

3. Functional Similarity

Amino acid or nucleotides with the same function are in the same column. No automatic program uses this criterion.

4. Sequences Similarity

Amino acids in the same column are those yield an alignment with maximum similarity most programs use this because it is the easiest criterion when the sequence is closely related. Structural, evolutionary and functional similarities are equivalent to sequence similarity.

The first three criterions have a clear biological meaning the fourth one does not. Yet when the sequences are similar enough, We can use similarity to place a multiple alignment that reflects the evolutionary, the structural and the functional relationships that exists between your sequences.

Main Applications of MSA

1. Extrapolation

A good MA can help to uncharacterized sequence is really a member of a protein family.

2. Phylogenetic Analysis

If we carefully choose the sequences to include in our MA, we can reconstruct the history of these proteins.

3. Pattern Identification

By discovering vs. conserved position, you can identify a region that is characteristic of a function (in protein or in nucleic acid sequence).

4. Domain Identification

It is possible to turn a MSA into a profile that describes a protein family or protein domain we can use this profile to scan databases for new members of the family.

Alignment with T-Coffee

T-coffee is the most recently developed methods for conducting multiple sequence alignments. It uses a principle that's a bit similar clustalW but yields more accurate alignments at the cost of slightly longer running time. T-coffee builds progressive alignments like clustalW, but it compare segment across the entire sequence set.

Possible Usage of T-Coffee Server

Usage	Description
Multiple Alignment	Given a set of sequence, T-Coffee can produce a multiple sequence alignment.
Evaluation	Evaluate the reliability of an existing multiple alignment
Using Structures	If same of our sequence have a known structure. T-Coffee can use them to help the alignment.
Combining Alignments	If you have several alignments of the same sequence produced with different methods (clustalW 1/ W and T-Coffee). T-Coffee to combine these alignments into single one.

BLAST - A Sequence Alignment Tool

Two sequences placed me below the other so that the residues match in corresponding position is known as alignment.

Global alignment : If the sequences match end to end we call it as a global alignment.

Local alignment : A small protein in the middle of the sequences alone is in alignment. Remaining residues are not in match. This kind of alignment is known as local alignment.

K K G A E Q E S V K E F L A K
K W A A E Q E S V K G N P A D

Blast is local sequence alignment tool.

Global Alignment Vs Local Alignment

Global alignment	Local alignment
Overall alignment	Relatively conserved short stretches
Stretches of low similarity	Unconserved region do not contribute to the measure of similarity
Preferred for direct comparison of two or more but limit number of sequences	Preferred for database search

Blastn : nucleotide - nucleotide BLAST

Blastp : protein - protein BLAST

Blastx : nucleotide query Vs protein database

tBlastn : protein query Vs protein database

tBlastx : nucleotide query Vs translated database

PSI-Blast : Position specific iterative Blast detect weak homologus by building a profile from a multiple alignment of the highest scoring hits in an initial BLAST search.

PHI-Blast : Pattern Hit Initiated Blast Combines Matching of regular expressions with local alignment surrounding the match.

Special → to compare two sequence of your choice.

1) Align two sequence (b/2 seq)

Megablast → For every large sequence

Blast Output

The Graphic Display

Shows you where your query is similar to other sequence. Depending on the server we use, this display can change a lot. It may also be absent on same server.

Red bar indicate similar sequence.

Pink bar indicate bit less good.

Green bar indicate not impressive at all.

Black bar indicate the worst scores and blue.

Red, pink and green matches are usually good ones.

Black hits are bad hits.

The hit list. The name of sequences similarly to our query, ranked by similarity.

Four Important Features

1. *The sequence accession number and the name :* This hyperlink takes you to the database entry that contains this sequence. Very important annotated information describing the query. A SWISS-PROT link (ISP may have useful information)

2. *Description* : A description that comes from the annotation.
3. *The bit score* : A measure of statistical significance of the alignment. The higher the bit score, the more similar the two sequences. Matches below 50 bits are very unreliable.
4. *The E-value (the expectation value)* : E-value provides the most important measure of statistical significance. Lower E-value, the more similar the sequence. The sequences identical to the query have an E-value very close to 0. Matches above 0.001 are often close to zone.

Alignments : Every alignment between the query and the reported hits. They have the following features:

The percent identity : This gives you a concrete substitute for the E-value. More than 25 percent identity is good news. The positives field gives a measure of fraction of residues that are either identical or similar - represent with a + on the actual alignment. The Gaps field shows the residues that were not aligned.

Score = 308 bits (789) Expect = 20-83

Identities = 203/381 (53%), Positive = 267/381 (69%)

Gaps = 22/381 (5%)

1. *Length* : This is the length of the alignment which indicates how long two segments of our sequences that BLAST has aligned are. It is the best way to see whether a hit means something. Sometimes very short alignments can come up with the high E-value and not be very meaningful.
2. *The top sequence* : Our query.
3. *The bottom sequence* : The hit or the subject.
4. *The line in between the sequence* : Between the two sequences is a line that contains a (+) sign for similar amino acids, a letter for identical residues or a space for mismatches.
5. *The xxxxx region* : In our query, BLAST makes automatically the regions with many low - complexity segments. They can cause troubles in the search. The Xs tell BLAST marking occurs in query sequence.
6. *The number* : Number to the side of the sequences indicates the coordinate of the match on our query sequence and on the hit

sequence. BLAST makes local alignments that may contain only a portion of the query and the hit.

7. *Query* : 280 KRKKEMTXXXXXXEGSEPFIGNSNDLAL 339
   KRKKEM + + FIGNS + L +
   209 KRKKEM – PKTIPEASFIGNSKEELKNT 264.

Parameters : A list of the various parameters used for the search

They include

1. The database
2. The BLAST program itself
3. The default parameters of the server.

In red life we have no control over this parameter. They always end up changing with time.

EMBL

Europe's primary nucleotide sequence resources. Main source for DNA and RNA sequences are direct submission from individual researchers, genome sequencing projects and patent application. The database is produced in collaboration with GeneBank and the DNA database of Japan (DDBJ). Each of the three groups collects a protein of the total sequence data reported worldwide and all new and updated database entries exchanged between the groups as a daily basis.

General Information

Primary Accession → AJ252272

Sequence retrieval system → System used at the ESI to search

SRS Entry ID → Form only EMBL: RJU252272.

Molecular type - Linear genomic DNA

Sequence length - 1594

Sequence version - AJ252272

Creation Date - 06 - Nov - 2000

Modification Date - 04 - Aug - 2004

Description - Rice tungro

Keywords - ORF3

Organism - Rice tungro bacilliform virus

Organism classification - Viruses

Reference - Reference given by a person features.

Key	Location	Qualifier	Value
Source	1...1594	Organism isolate mol-type country	Rice tungro bacilliform virus Andhra Pradesh genomic DNA
Key	Location	Qualifier	Value
Cds	<1...>1591	Codon start Product Function Note db_Xref Protein - id	1 Polyprotein Movement Protein and ORF3 Coat protein Uniprot tkB/ TREMBL CAC15488 Translation
Sequence Characteristic		length, A count, C count, T count, other count, G count	

SWISS – PROT

Swiss - Prot is a manually curated biological database of protein sequences. Swiss Prot was created in 1986 by Amos Bairoch during his PhD and developed by Swiss Institute of Bioinformatics and the European Bioinformatics Institute. Swiss - Prot strives to provide reliable protein sequences associated with high level of annotation (such as the description of the function of a protein its domains structure post translational modification, variant etc).

In 2002, the Uniprot consortium was created; it is collaboration between the Swiss Institute of Bioinformatics, the European Bioinformatics Institute and the protein Information Resources (PIR) to produce the Uniprot knowledge base, the world's most comprehensive catalogue of information on proteins. UniProt knowledge base (UniProt) is the central access point for extensive curated protein information, including function, classification and cross reference.

TrEMBL, Translated EMBL

TrEMBL, Translated EMBL is a very large protein database is Swiss - Prot format generlated by computer translation of genetic information from the EMBL database.

Computer translation is not entirely perfect, so proteins predicated by the TrEMBL database and be hypothetical and many TrEMBL entries are poorly annotated.

TrEMBL is current being combined with the above two database in the UniProt project.

General Information About the Entry

Entry Name : SWISS - PROT identifier gives a quick idea of what entry is all about. Fly lysozyme entry name is LYSA - DROME, certainly more ambiguous and more difficult to remember (DROME Stands for DROSOphila MElanogaster)

Primary Accession Number : P00533 - This is the truly unique and stable identifier of this sequences. Accession number are great because they allow you to retrace the database history of a given protein.

Secondary Accession Numbers : P06268, 214225 secondary accession numbers make sure that when you read an old paper mentioning Q12225 or P06268 as accession number. Whatever secondary accession number you type in, you always fall back on the **current one** : P00533.

Entered in SWISS - PROT : Release 01, July 1986.

Sequence was last modified in : Release 35, November 1997.

Annotations were last modified in : Release 41, June 2002.

Name and origin of the Protein.

Protein name : Epidermal growth factor receptor : This field contains general descriptive information about the sequence. It is generally sufficient to identify the protein precisely.

Synonyms: Receptor protein - tyrosine kinase, EC 2.7.1.112:

The E.C. number (2.7.1.112) encodes the biochemical reaction that this protein performs. E.E. stands for Enzyme Nomenclature committee.

Gene name : EGFR or ERBB1: This field is useful when you need to interact with nucleotide sequence and genome databases.

From : Homo sapiens (Human) [Tax ID: 9606] : The from field defines precisely the origin of the protein.

Taxonomy : Eukaryota: Metazoa: chordata, craniate, vertebrate: This field lets the taxonomic classification of the source organism. The taxonomic classification used to maintain at NCBI.

Cross Reference

The Cross-References section contains links to entries in other databases that contain some information about our protein.

The major areas are

EMBL : Nucleotide sequence world.

PIR : Protein Information Resources has Accession numbers.

HSSP : Link to a sequence homologue of current query for which 3-D structural information is avaliable

SWISS-2DPAGE : Collection of 2-D gel electrophoreses for different human tissues.

GENEW : For human protein only.

MIM : Mostly for human / mouse protein.

Gene Cards : Gene card database.

Gene Lynx : Gene Lynx is collection of hyperlinks to publicly available resources associated with each human gene.

GlycosuiteDb : Link to chemical position of sugar moiety.

Source : This field is only for human proteins.

Ensembly : Only human and mouse protein.

Profiles/domain : Different algorithm and detection of recurrent structural/sequence domains, as well as for clustering of proteins in families.

DIP : Database of Interacting proteins of database.

Key Words

Simple list of terms relevant to your current protein, such as transmembrane, receptor, glycoprotein.

Features

Information on the protein is precisely mapped onto the sequence.

Key from to length description.

Key field → Features looking for the numbers in the From and To

correspond to N-terminus → C-terminus numbering of the sequence.

length - length of the sequences.

Description - well a description.

Key Terms

1. Signal → Residues 1 to 24 are a signal peptide.
2. Chain - Residues 25 to 1210
3. Domain - Residues 25 to 645 constitute the part of the mature protein.
4. Transmem → 23 residues (646 to 668).
5. Repeat → 75-300 segment is similar to the 390-600 segment.

Sequence

Sequence with FASTA format

Rice Genome Project

Rice is one of the main staples in the world and is cultivated mainly in Asia, Africa, and Latin America using irrigated and upland ground for paddy fields. The major limiting factors for rice cultivation are climate, soil, and water. Except for the upland rice in Africa, rice plants grow mainly in tropical and semitropical areas characterized by a rainy season. Thc wild relatives of cultivated rice, *Oryza sativa*, are also widely distributed within nearly 20° north and south latitude (IRRI 2004).

The domestication of rice is undoubtedly one of the most important developments in human history and, at present, about half of the world population relies on rice for its daily main food source. Despite the long history of breeding and improvement of rice varieties adapted to a wide range of environmental conditions as well as varieties highly preferred by both farmers and consumers, novel rice varieties are still necessary to meet the demands of increasing population pressure coupled with decreasing resources for rice production.

What is Genomics?

- The genome of a plant, animal or microbe is the totality of its genetic information including all the genes and the non-transcribed sections of the DNA.
- Genomics is the study of this complete genetic content of an organism.

Model Experimental Crop : Why Rice?

- Belongs to the grass family
- Have a much smaller genome and the greatest biodiversity of cereal crops.
- Has a high degree of colinearity with the genomes of wheat, barley, and maize.

Genome sizes	
Species	No. of base pairs
Arabidopsis	125 million
Drosophila	180 million
Rice	400 million
Maize	2,500 million
Human	3,000 million
Barley	4,900 million
Wheat	16,000 million

Why Rice Genome was Sequenced?

- It can address many different aspects of rice research, including genetic diversity and productivity improvement.
- To design efficient ways to tap into the wealth of rice genome sequence information to address production constraints in an environmentally sustainable manner.

Milestones in the Rice Genome Sequencing

Sept 1997 : Sequencing of the rice genome was initiated as an international collaboration among 10 countries.

Feb 1998 : IRGSP (International Rice Genome Sequencing Project) was launched under the coordination of the Rice Genome Program (RGP) of Japan.

April 2000 : Monsanto Co. produced a draft sequence of BAC contigs covering 260 Mb of the rice genome; 95% of rice genes were identified.

Feb 2001 : Syngenta produced a draft sequence and identified 32,000 to 50,000 genes, w/ 99.8% sequence accuracy and identified. 99% of the rice genes.

Dec 2002 : IRGSP finished high-quality draft sequence (clone-by-clone approach) with a sequence length, excluding overlaps, of 366 Mb corresponding to ~92% of the rice genome.

Dec 2004 : IRGSP produce the high-quality' sequence of the entire rice genome; with 99.99% accuracy and without any sequence gap.

How Rice was Sequenced?

Map-based sequencing - each stretch of sequence is physically anchored to a chromosome forming clones, which are individually sequenced.

Applications and Impact

- Development of gene- specific markers for marker-assisted breeding of new and improved rice varieties
- Understand how a plant responds to the environment and which genes control various functions of the plant
- Improve the nutritional value of rice, enhance crop yield by improving seed quality, resistance to pests and diseases and plant hardiness
- Useful in identifying plant-specific genes that can be potential herbicide targets.

Benefits

- Comparative genomics - Comparisons across species will be useful in understanding the basis of major biological processes
- Using the high quality rice genome sequencing is a good standard for evaluating the gene coverage.
- Hasten the development of genetic markers in rice as well as in other grass genomes by focusing only in regions of the genome containing the gene of economic interest or importance.
- The rice genome holds fundamental information in its biological "power", including physiology, development, genetics and molecular evolution.

  Ssequence BLAST using TIGR (*http://www.tigr.org/tdb/e2k1/osa1*)
- Downloading rice protein sequences from TIGR (*http://www.tigr.org/tdb/e2k1/osa1/LocusNameSearch.shtml*)

Software for Bio-informatics

The use of software in biological applications has given a new dimension to the field of bioinformatics. Freeware software is available on the internet and many universities have established servers. There are number of simple stand-alone programs as well as complex integrated network version of the commercial software available. Factors that must be taken into consideration when designing bioinformatics tools, software and programmes are:

1. Software should be accessible to all, easy to use and should run on normal PC Maximum concentrated on GUI interface and to some extent DOS commands to avoid command interface.
2. Software list selected below to run some of common tasks in bioinformatics, which should be usable to person from varied fields from life sciences/molecular biology background.
3. Last but not the least, software's should be easy to install without complex commands.

Criteria

The criteria for bioinformatics software:

- Software should be accessible to all, easy to use and should run on normal PC.
- Maximum concentration should be given on GUI interface and to some extent DOS commands to avoid command interface.
- Software should be usable to person from varied fields from life sciences/molecular biology background.
- Software's should be easy to install without complex commands.

Prerequisite

Familiarity to use Operating system and to know unzip or to run exe files. The most commonly used free software are RASMOL, CHIME, MOLMOL and protein explorer.

Rasmol – Offline Use

Rasmol is a molecular visualization program use for studying protein and DNA structures. This software was developed by Roger satyle. This software is downloaded from.

http://www.umass.edu/microbio/rasd.

It runs in the operating system such as Microsoft Windows, Unix. The program reads in a molecule coordinate file and interactively displays the molecule on the screen in a variety of color schemes and molecular representations. The displayed molecule may be rotated, translated, zoomed using the mouse, the scroll bars, or a dialog box.

Various menu options available in Rasmol

1. File
2. Edit
3. Display
4. Color
5. Options
6. Export
7. Help

CHIME

Centre online for Health Information and MultiProfessional Education CHIME is a molecular visualization tool, which is a remote server based program in sharp contrast to the stand alone software Rasmol, Chime only in-line, when connected to the internet. Rasmol can be used offline, when not connected to the internet.

www.umass.edu/microbio/chime

Chime is partly derived from Rasmol and it program code is proprietary whereas Rasmol code is public.

MOLMOL

Download and use it Molecule analysis and Molecule display.

ftp.mol.biol.eth.ch

MOLMOL is a molecular graphics program for display, analysis and manipulation of three dimensional structures of biological macromolecules for the solution structure of protein and nucleic acids protein explorer (advanced version of rasmol) A moecular visualization tools like Swiss PDB viewer.

GCG

The Genetic Computer Group (GCG) company was one of the pioneers of bioinformatics and its Wisconsin package tools are widely used for sequence analysis such as sequence alignment, secondary structure prediction, restriction site analysis, repeat finder, primer design, similarity searches etc.

SPDBV

SPDBV can be downloaded as follows:

SPDBV ..... Swiss PDB viewer → Deep view

website : http : //us.expasy.org/spdbv/

SPDBV is also routinely used for structure comparison, mutation, modeling and geometry analysis like bound length, bound angle etc.

Options in SPDBV

1. File
2. Edit
3. Select
4. Build
5. Tools
6. Fit
7. Display
8. Color
9. Preferences
10. Swiss Model
11. Window
12. Help

Major Categories of Bioinformatics Software

Current categories for the Bioinformatics Software Section are:

1. Sequence Databases
2. Pathway Analysis Tools
3. Structure Prediction and Analysis Tools
4. Sequence Analysis Tools
5. Sequence Management Tools
6. Visualization Tools

1. Sequence Database

GENESEQ was a database of protein and nucleic acid sequences extracted from world-wide patent documents. The program was offered Oxford Molecular Group but is no longer available.

2. Pathway Analysis Tools

BIOCARTA is a web-based, user contributed collection of pathways and the tools to study them. Observe how genes interact in dynamic graphical models. The online maps depict molecular relationships from areas of active research. In an "open source" approach, this community-fed forum constantly integrates emerging proteomic information from the scientific community. It also catalogs and summarizes important resources providing information for over 120,000 genes from multiple species. Find both classical pathways as well as current suggestions for new pathways.

3. Structure Prediction and Analysis Tools

FGENEH from the Baylor College of Medicine predicts a single complete gene and the corresponding amino acid sequence.

4. Sequence Analysis Tools

GenomeBrowser is a powerful software tool that simplifies the proccess of analysis, annotation, and manipulation of genetic sequences. The underlying principle of UniPro GenomeBrowser is an interactive visualization of genetic sequences with ability to navigate, zoom and create annotations for selected parts of sequences.

Advanced search tool, powerful contextual statistics, support of major genetic data formats and collection of pluggable modules increase the efficiency, simplify and in a number of cases, make the process of analysis automatic. Final results as well as intermediate ones are graphically presented and can be saved and used in a future analysis.

5. Sequence Management Tools

BioMerge Developed by Molecular Informatics, Inc. (MII) and marketed in partnership with PE Biosystems, Inc. BioMerge is a complete system for managing biological sequence data and related information. Fast becoming the industry standard for managing sequence data, this bioinformatics tool brings together public, third-party and proprietary sequence data into a single, fully functional relational database management system.

6. Visualization Tools

KoriBlast is a reliable graphical environment dedicated to Blast data mining. KoriBlast combines Blast searches with advanced data management capabilities and a state-of-the-art graphical user interface.

Internet Sites for Bioinformatics

Name	Address	Description
GeneBank	*www.ncbi.nlm.nih.gov*	DNA database.
EMBL	*www.ebi.ac.uk/embl*	Nucleotide sequence
Swiss-Prot	*www.expasy.ch*	Annotated protein sequence
PDB	*www.rcsb.org*	Protein structure
Blast	*www.ncbi.nlm.nih.gov*	Database search

Example I - Search the Protein Structure for the Crop Rice

Protein Data Bank

The RCSB (Research Collaboratory for Structural Bioinformatics) PDB provides a variety of tools and resources for studying the structures of biological macromolecules and their relationships to sequence, function and disease.

Step 1: Point the browser at *www.rcsb.org*

Step 2: Give the crop name rice in search box

Step 3 : Click the first heading to get the structure of rice.

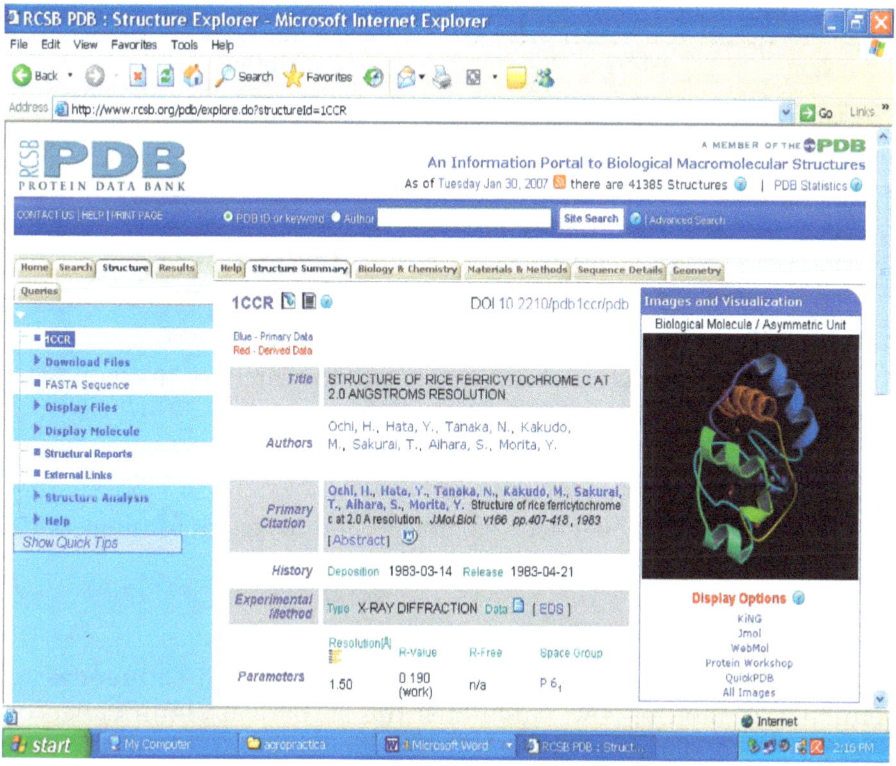

Step 4 : Click the image to get the full view.

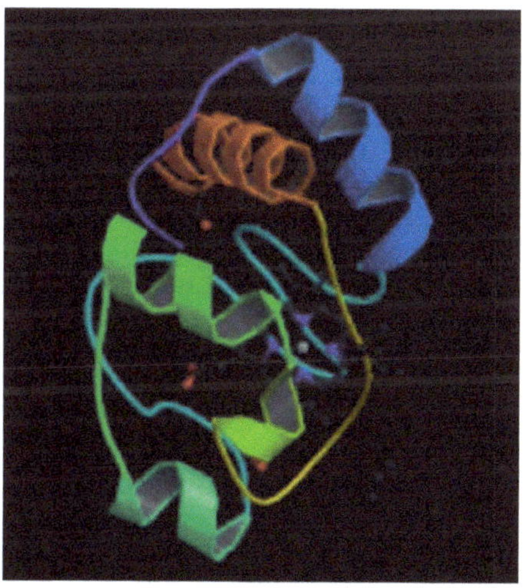

Structure of rice

Exercise to Practice

1. Web references

Name	Address	Description
GeneBank	www.ncbi.nlm.nih.gov	DNA database.
EMBL	www.ebi.ac.uk/embl	Nucleotide sequence
Swiss-Prot	www.expasy.ch	Annotated protein sequence
PDB	www.rcsb.org	Protein structure
Blast	www.ncbi.nlm.nih.gov	Database search

Example II - Similarity Search on Sequence Database

Find similarity between sequences and to search for nucleotide or protein sequences in sequence database for rice

Aim

To find similarity between sequences and to search for nucleotide or protein sequences in sequence database

BLAST (Basic Logical Alignment Search Tool)

This program compares nucleotide or protein sequences to sequence databases and calculates the statistical significance of matches. BLAST can be used to infer functional and evolutionary relationships between sequences as well as help identify members of gene families.

Types of BLAST

Protein

Blastp - compares a protein sequence with a protein database.

Nucleotide

tblastn - compares a protein sequence with a nucleotide database.

Megablast - search for highly similar sequences.

Steps involved

- Browse to www.ncbi.nlm.nih.gov/BLAST
- Access sequence from database eg: gene in rice.
- Search the similarity by using different types of blast.

BLAST output

- Graphic display - query is similar to other sequence.
- Hit list - name of sequence similar to query ranked by similarity. Showing sequence number and the name.
- Bit score - a measure of the statistical significance of the alignment.
- E-Value (expectation value) - measure similarity and homology.
- Alignment - every alignment between the query and report the list percent identity and length.

FASTA format

A sequence in FASTA format begins with a single-line description, followed by lines of sequence data. The description line is distinguished from the sequence data by a greater than (">") symbol in the first column. It is recommended that all lines of text be shorter than 80 characters in length.

Amino acid official code

A alanine
B aspartate
C cystine
D aspartate
E glutamate
F phenylalanine
G glycine
H histidine
I isoleucine
K lysine
L leucine
M methionine
N asparagine
P proline
Q glutamine
R arginine
S serine
T threonine
U selenocysteine
V valine
W tryptophan
Y tyrosine
Z glutamate or glutamine
X any
* translation stop
- gap of indeterminate length

Retrieved Protein Sequence from Swiss-Protein

>Q84J55|14331_ORYSA 14-3-3-like protein GF14-A - Oryza sativa (Rice).

MAAAAGGGTREEMVYMAKLAEQAERYEEMVEFM EKVVTAAAAGGGGELTVEERNLLSVAYKNVIG ARRASWRI

VSSIEQKEEGRGAAGHA AAARSYRAR VEAELSNICAGILRLLD ERLVPAAAAVDAKVFYLKMKGDYHRYLAEFKTG AERKD AADAT LAAYQA AQDIAMKELSPTHPI

RLGLALNFSVFYYEILNSPDRACTLAKQAFDEA ISELDTLGEESYKDSTLIMQLLRDNLTLWTSDMQDDGGD EMRDATKPE DEH

Using BLAST to compare our protein sequence to other protein sequence

1. Point the internet browser to www.ncbi.nlm.nih.gov/BLAST
2. Click the Standard Protein-Protein BLAST link
3. Open the enzyme FASTA formatted protein sequence file.
4. Using your browser's Edit menu, copy and paste one of the protein sequences into the BLAST search window.
5. Deselect the Do CD-search, but don't change the choose database setting.
6. Click BLAST button.

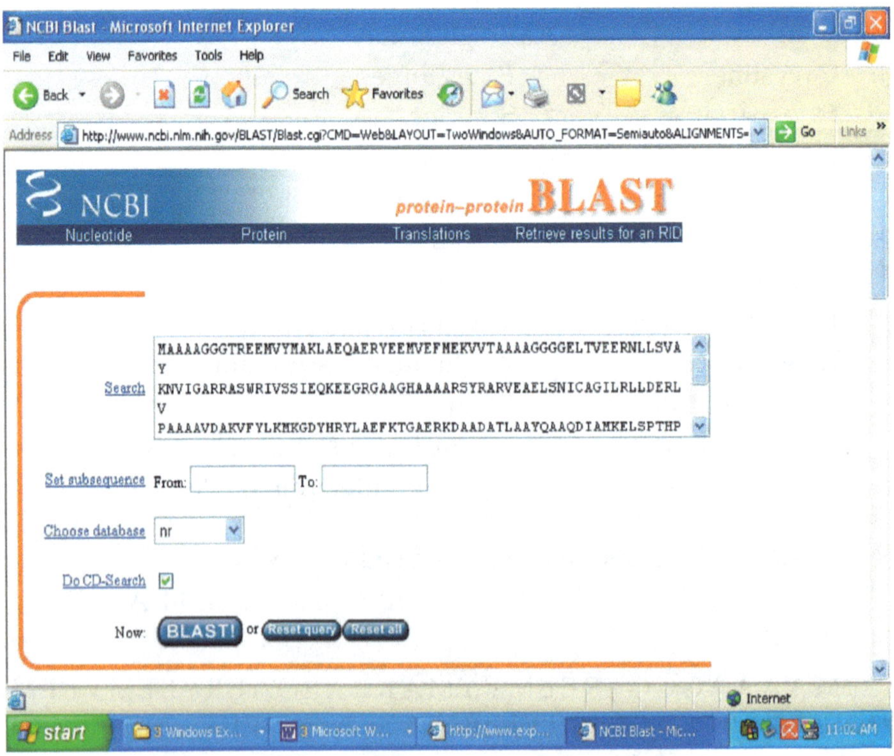

Formatting BLAST

Your request has been successfully submitted and put into the Blast Queue.

Query = (264 letters)

Putative conserved domains have been detected, click on the image below for detailed results.

The request ID is 1165210329-30727-11812527919.BLASTQ4

Format! or Reset all

Result of BLAST

Query=

Length=264

Graphical Display

Distribution of 100 Blast Hits on the Query Sequence

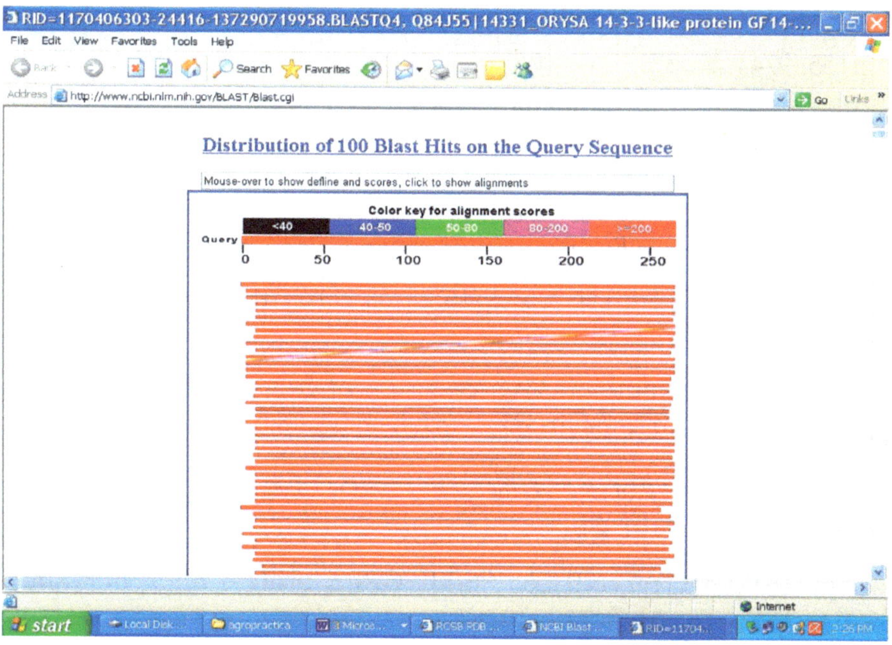

Text Display

Sequences producing significant alignments: (Bits) Value

gi|115476928|ref|NP_001062060.1| Os08g0480800 [Oryza sativa (G 421 2e-116

gi|9798605|dbj|BAB11740.1| TaWIN2 [Triticum aestivum] 390 4e-107 U G

gi|1168189|sp|P42653|1433A_VICFA 14-3-3-like protein A (VFA-1... 375 1e-102

gi|2879818|emb|CAA72382.1| 14-3-3 protein [Solanum tuberosum] 375 1e-102 U

gi|4775555|emb|CAB42547.1| 14-3-3-like protein [Pisum sativum] 375 1e-102

gi|13928452|dbj|BAB47119.1| 14-3-3 protein [Vigna angularis] 375 2e-102

gi|42491254|dbj|BAD10939.1| 14-3-3 protein [Nicotiana tabacum... 374 2e-102

gi|3912951|sp|O49998|1433F_TOBAC 14-3-3-like protein F >gi|26... 374 2e-102

gi|1168196|sp|P46266|1433_PEA 14-3-3-like protein >gi|555974|... 373 6e-102

gi|26454608|sp|P93211|14336_LYCES 14-3-3 protein 6 >gi|156371... 372 6e-102 G

gi|1168195|sp|P29307|1433_OENHO 14-3-3-like protein >gi|20062... 372 8e-102

gi|4850247|emb|CAB42546.2| 14-3-3-like protein [Pisum sativum] 372 1e-101

gi|3766535|emb|CAA72383.1| 14-3-3 protein [Solanum tuberosum] 372 1e-101 U

gi|44917143|dbj|BAD12172.1| 14-3-3 c-2 protein [Nicotiana tabacu 372 1e-101 U

gi|18411901|ref|NP_565176.1| GRF2; protein phosphorylated ami... 372 1e-101 U G gi | 487791 | gb | AAA96253.1 | GF14omega isoform 371 1e-101

gi | 44917155 | dbj | BAD12178.1 | 14-3-3 f-1 protein [Nicotiana tabacu 371 2e-101 U

gi | 3023189 | sp | P93343 | 1433C_TOBAC 14-3-3-like protein C (14-3-... 371 2e-101

gi | 47558857 | gb | AAT35546.1 | 14-3-3 protein [Tropaeolum majus] 371 2e-101

gi | 13447104 | gb | AAK26634.1 | AF342780_1 GF14 omega [Brassica napus] 370 2e-101 U

gi | 5902676 | sp | O65352 | 1433_HELAN 14-3-3-like protein >gi | 31539... 370 3e-101

gi | 23558747 | emb | CAC84142.3 | 14-3-3 protein [Nicotiana tabacum] 370 3e-101 U

gi | 3912948 | sp | O49995 | 1433B_TOBAC 14-3-3-like protein B >gi | 26... 370 5e-101

gi | 3023190 | sp | P93784 | 14335_SOLTU 14-3-3-like protein 16R >gi | ... 369 8e-101

Seqeuence Alignment

>gi | 115476928 | ref | NP_001062060.1 | G Os08g0480800 [Oryza sativa (japonica cultivar-group)]

gi | 75297659 | sp | Q84J55 | 14331_ORYSA 14-3-3-like protein GF14-A (G-box factor 14-3-3 homolog A)

gi | 29367361 | gb | AAO72553.1 | U G WIN2-like protein [Oryza sativa (japonica cultivar-group)]

gi | 29367565 | gb | AAO72644.1 | U G TaWIN2-like protein [Oryza sativa (japonica cultivar-group)]

gi | 42408588 | dbj | BAD09765.1 | G putative TaWIN2 [Oryza sativa (japonica cultivar-group)]

gi | 42409022 | dbj | BAD10275.1 | G putative TaWIN2 [Oryza sativa (japonica cultivar-group)] gi|113624029|dbj|BAF23974.1| G Os08g0480800 [Oryza sativa (japonica cultivar-group)]

Length=264

Score = 421 bits (1081), Expect = 2e-116, Method: Composition-based stats.

Identities = 264/264 (100%), Positives = 264/264 (100%), Gaps = 0/264 (0%)

Query 1

MAAAAGGGTREEMVYMAKLAEQAERYEEMVEFMEKVVTAAAAGG GGELTVEERNLLSVAY 60

MAAAAGGGTREEMVYMAKLAEQAERYEEMVEFMEKVVTA AAAGGGGELTVEERNLLSVAY

Sbjct 1

MAAAAGGGTREEMVYMAKLAEQAERYE EMVEF MEKV VTA AAAGGGGELTVEERNLLSVAY 60

Query 61

KNVIGARRASWRIVSSIEQKEEGRGAAGHAAAARS YRARVEAEL SNICAGILRLLDERLV 120

KNVIGARRASWRIVSSIEQKEEGRGAAGHAAAA RSYRARVE AELSNICAGILRLLDERLV

Sbjct 61

KNVIGARRASWRIVSSIEQKEEGRGAAGHAAAARS YRARVEAELSNICAGILRLLDERLV 120

Query 121

PAAAAVDAKVFYLKMKGDYHRYLAEFKTGAERKD AADATL AAYQAAQDIAMKELSPTHPI 180

PAAAAVDAKVFYLKMKGDYHRYLAEFKTGA ERKDAADAT LAAYQAAQDIAMKELSPTHPI

Sbjct 121

PAAAAVDAKVFYLKMKGDYHRYLAEFKTGAERKDAAD ATLAAYQAAQDIAM KELSPTHPI 180

Query 181

RLGLALNFSVFYYEILNSPDRACTLAKQAFDEAISELD TLGEESY KDSTLIMQLLRDNLT 240

RLGLALNFSVFYYEILNSPDRACTLAKQAFDEAISELDT LGEESYKDSTLIMQLLRDNLT

Sbjct 181

RLGLALNFSVFYYEILNSPDRACTLAKQAFDEA ISELDTL GEESYKDSTLIMQLLRDNLT 240

Result

The lower the e value of the search higher is the similarity between the two sequence. For Q84J55|14331_ORYSA 14-3-3-like protein GF14-A - Oryza sativa (Rice) the similar protein id is gi|115476928| ref|NP_001062060.1| in protein blast.

Exercise to Practice

1. Find similarity between sequences and to search for nucleotide or protein sequences in sequence database for rice.
2. Find similarity between sequences and to search for nucleotide or protein sequences in sequence database for enyzme.
3. Find similarity between sequences and to search for nucleotide or protein sequences in sequence database for wheat.

Example III - Searches Using SWISS-PROT, GENEBANK and EMBL

Search the protein and nucleotide sequence.

Aim

To characterize protein and nucleotide sequence

Protein Sequence using SWISSPROT – Protein Sequence for Rice

Entry information	
Entry name	**14331_ORYSA**
Primary accession number	**Q84J55**
Secondary accession numbers	None
Integrated into Swiss-Prot on	July 11, 2006
Sequence was last modified on	June 1, 2003 (Sequence version 1)
Annotations were last modified on	November 28, 2006 (Entry version 19)
Name and origin of the protein	
Protein name	**14-3-3-like protein GF14-A**
Synonym	**G-box factor 14-3-3 homolog A**
Gene name	**Name: GF14A** OrderedLocusNames: Os08g0480800, LOC_Os08g37490 ORFNames: OJ1113_A10.40, OSJNBb0092C08.10
From	Oryza sativa (Rice) [TaxID: 4530]
Taxonomy	Eukaryota; Viridiplantae; Streptophyta; Embryophyta; Tracheophyta; Spermatophyta; Magnoliophyta; Liliopsida; Poales; Poaceae; BEP clade; Ehrhartoideae; Oryzeae; Oryza.

264 AA [This is the length of the unprocessed precursor]	Molecular weight : **29001 Da** [This is the MW of the unprocessed precursor]	CRC64: **F4094830FEF61600** [This is a checksum on the sequence]

10 20 30 40 50 60 MAAAAGGGTR EEM VYMAKLA EQAERYEEMV EFMEKVVTAA AAGGGGELTV EERNLLSVAY 70 80 90 100 110 120 KNVIGARRAS WRIVSSIEQK EEGRGAAGHA AAARSYRARV EAELSNICAG ILRLLDERLV 130 140 150 160 170 180 PAAAAVDAKV FYLKMKGDYH RYLAEFKTGA ERKDAADATL AAYQAAQDIA MKELSPTHPI 190 200 210 220 230 240 RLGLALNFSV FYYEILNSPD RACTLAKQAF DEAISELDTL GEESYKDSTL IMQLLRDNLT 250 260 LWTSDMQDDG GDEMRDATKP EDEH	Q84J55 in FASTA format

FASTA format

>Q84J55|14331_ORYSA 14-3-3-like protein GF14-A - Oryza sativa (Rice).

MAAAAGGGTREEMVYMAKLAEQAERYEEMVEFMEKV VTAAAAGGGGELTVEERNLLSVAYKNVIGARRASWRIVSSIEQKEEGR GAAGHAAAARSYRARVEAELSNICAGILRLLDERLVPAAAAVDAKVFYL KMKGDYHRYLAEFKTGAERKDAADATLAAYQAAQDIAMKELSPTHPI

RLGLALNFSVFYYEILNSPDRACTLAKQAFDEAISELDTL GEESYK DSTLIMQLLRDNLTLWTSDMQDDGGDEMRDATKPEDEH

GENEBANK

Nucleotide Sequence for Rice using GeneBank

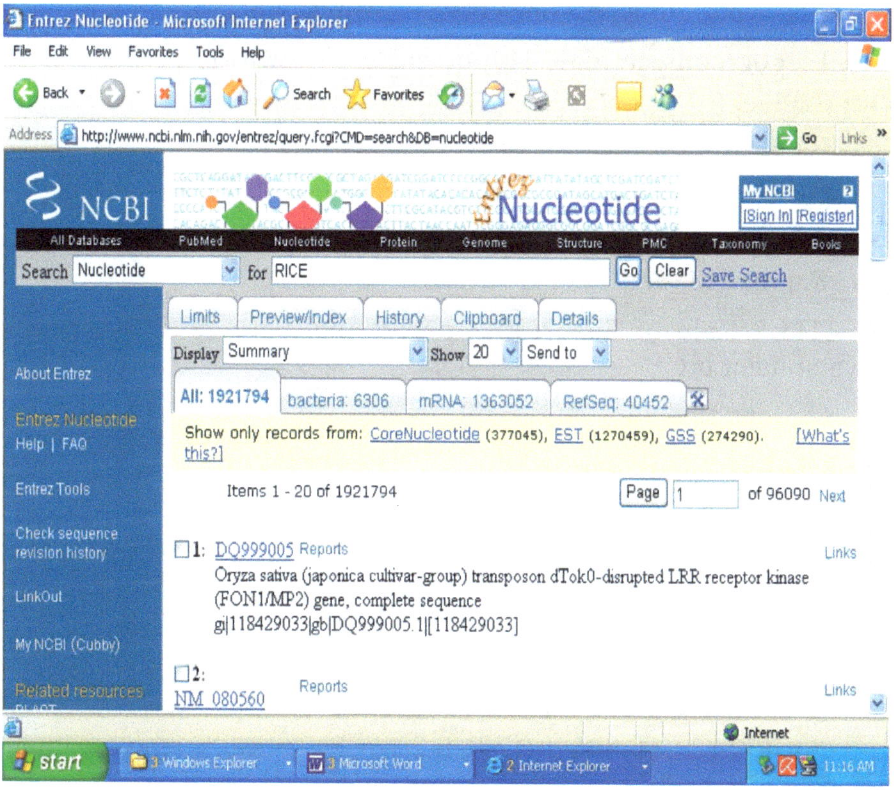

1: DQ855948. Fusarium equiseti...[gi:116833059]	

Features Sequence

1: DQ855948. Fusarium equiseti...[gi:116833059]

Features Sequence

LOCUS DQ855948 598 bp DNA linear PLN 01-DEC-2006

DEFINITION Fusarium equiseti isolate DAOM170306 translation elongation factor

1 alpha (EF-1 alpha) gene, partial sequence.

ACCESSION DQ855948

VERSION DQ855948.1 GI:116833059

KEYWORDS SOURCE Fusarium equiseti

ORGANISM Fusarium equiseti

Eukaryota; Fungi; Ascomycota; Pezizomycotina; Sordariomycetes;

Hypocreomycetidae; Hypocreales; mitosporic Hypocreales; Fusarium.

REFERENCE 1 (bases 1 to 598)

AUTHORS Punja,Z.K., Wan,A., Goswami,R.S., Barasubiye,T., Seifert,K.A.,

Rahman,M. and Levesque,C.A.

TITLE Role of Fusarium Species in Rusty Root Symptom Development on

Ginseng Roots

JOURNAL Unpublished

REFERENCE 2 (bases 1 to 598)

AUTHORS Punja,Z.K., Wan,A., Goswami,R.S., Rahman,M., Barasubiye,T.,

Seifert,K.A. and Levesque,C.A.

TITLE Direct Submission

JOURNAL Submitted (17-JUL-2006) Biological Sciences, Simon Fraser

University, 8888 University Drive, Burnaby, BC V5A 1S6, Canada

FEATURES Location/Qualifiers

source 1..598

/organism="Fusarium equiseti"

/mol_type="genomic DNA"

/isolate="DAOM170306"

/isolation_source="rice"

/db_xref="taxon:61235"

/country="India"

gene <1..>598

/gene="EF-1 alpha"

/note="translation elongation factor 1 alpha; coding region not determined"

ORIGIN

1 caagtcgacc actgtaagta caaccaacag cgggttgctt atctgcactc ggaatccgcc
61 aaacctggca gggtatcacc aaaacatctt gctaactttt gacagaccgg tcacttgatc
121 taccagtgcg gtggtatcga caagcgaacc atcgagaagt tcgagaaggt tagtcaatat
181 cccttcgatt acgcgcgctc ccatcgattc ccacgattcg ctccctcact cgaaacacat
241 ccattacccc gctcgagtcc gaaaattttg cggtgcgacc gtgatttttt ctggtggggt
301 atcttacccc gccactcgag tcacggatgc gcttgccctg ttcccacaaa accttaccac
361 cctgtcgcgc actacatgtc ttgcagtcac taaccactgg acaataggaa gccgccgagc
421 tcggaaaggg ttccttcaag tacgcctggg ttcttgacaa gctcaaagcc gagcgtgagc
481 gtggtatcac cattgatatc gctctctgga agttcgagac tcctcgctac tatgtcaccg
541 tcattggtat gttgtcactg tctcacacta tcatgtattc atcatgctaa catctctc

EMBL – Nucleotide Sequence

General Information	
Primary Accession #	AJ252272
Accession #	AJ252272
SRS Entry ID	EMBL:AJ252272 (*formerly EMBL:RTU252272*)
Molecule Type	linear genomic DNA
Sequence Length	1594
Entry Division	VRL (*Viruses*)
Entry Data Class	STD (*Standard*)
Sequence Version	AJ252272.1
Creation Date	06-NOV-2000
Modification Date	04-AUG-2004
Description	
Description	Rice tungro bacilliform virus partial ORF3
Keywords	ORF3.;
Organism	Rice tungro bacilliform virus
Organism Classification	Viruses; Retro -transcribing viruses; Caulimoviridae; Tungrovirus.
References	
1.	Venkataramana,M.; Submitted (06 -JAN-2000) to the EMBL/GenBank/DDBJ databases. Venkataramana M., Department of Virology, Sri Venkateswara University, Tirupati, Andhra Pradesh 517502, INDIA. Position 1-1594
2.	Venkataramana,M.; **Studies on tungro disease of rice in Andhra Pradesh: Characterisation of rice tungro spherical (RTSV) and rice tungro bacilliform (RTBV) viruses and development of tests for detection of RTBV.**

Features			
Key	**Location**	**Qualifier**	**Value**
source	1..1594	**organism**	Rice tungro bacilliform virus
		isolate	Andhra Pradesh (AP)
		mol_type	genomic DNA
		country	India
		virion	
		db_xref	taxon:10654
cds	<1..>1594	**codon_start**	1
		product	polyprotein
		function	movement protein and coat protein
		note	ORF3
		db_xref	UniProtKB/TrEMBL:Q9DUR8
		protein_id	CAC15488.1
		translation	

Sequence	
Characteristics	**Length:** 1594 BP, **A Count:** 699, **C Count:**232, **G Count:** 298, **T Count:**365, **Others Count:**0

Sequence

```
>embl|AJ252272|AJ252272 Rice tungro bacilliform virus partial ORF3
aagctttattaccatataggcatgatggccataggaattaagggattacatagaaagaaa
ataggtactaaagtaatggtgatgttttatgacgatagtttcggtaaagctaaagaagct
agtataggatcaatagaaatggatatgaatgcaggttgcggagtattttactcttgccct
gacttcgctaaatatataaaagacctatcccatcttaagataggaatacaaaccctagga
tatgaaaactatgagggaaaaaatctttccatagccataaaaactataggaagattaaca
actaatatccaggctagatataaaatgaatgttaaagatatagttgaacaaatatcgtcc
caaggaataactatggttgcacccatggaaatagatccatcacatttagatgggaatgaa
tggaatttaagcaaattcatgattcaagaaggaacaagtagagttcctaataaagcccta
atatatcagaatttgcatggaggagaatcgcttagattttcaaactatacgcaaaccaaa
atgcacgatccaacagaaataaattctgatgaagatgaagatttaaaaattctaggagaa
caattaaatgccaaaatggcaacctttcaagaagaaaccctagaacaaaaactagaacgc
ataaaagaagaaaagaaacaacttctagcaaaactagaagctaaaaagaaagaaatagaa
gcaaaatctttaaatatggcagtcatcgaagatgacttcaacccaaacaacgaatatcta
gatgattcatattctgaattagaagatctagaattcgaacaattaggattaactggttgg
gaagatctagaccaagaaagtctagaaacagaagaaattactgaatgggaaaacccaaac
caagtcctaaccagagaaataaaagcctttaaatcagtctctgaacaaatagaagatata
tttggagaactattaaaagaacatggaaattatgacatggcccttaagaatttagaagaa
agatatgatctaaaaaacttagaaggagccaaaaacctagaagaaatagccaaagcatcc
acatcaaaaatgatggatgtaaaaccagtaaaaagaccaaaagaagaacaaacagcatac
gaagatgatatgagggatgatgggagaagaaaagaattacaacccaatccagaagtatcc
tcaaaagatagaaactttgaaagaataggaggatcatataagaaaaacttttaccctagt
aaaagtgaaatcctaaacctagaccatgtacccccacaattttattacgaccaaataata
acctgggaaggaatagtaaaaaatgaatgggaagctagaaaaaggatggtatggacatg
tggtcttggatggatggaagaataacaggtatggttttatatttaatacaagattggata
tccaaaaaccaagccgcttataatgatataaaatcccgaggagatagaccagaaaatttt
gtaaaaatggtaaaagataggttcttaatagaagatcctacagatgaaaggagaacagcc
ttacaaagattagctctaagagaattagaagctt
```

Output

Protein Sequence using Swissprot for Rice

MAAAAGGGTREEMVYMAKLAEQAERYEEM
VEFMEKVVTAAAAGGGGELTVEERNLLSVAYKNV
IGARRASWRIVSSIEQKEEGRGAAGHAAAARSYRARVEAE
LSNICAGILRLLDERLVPAAAAVDAKVFYLKMKGDY
HRYLAEFKTGAERKDAADATLAAYQAAQ
DIAMKELSPTHPIRLGLALNFSVFYYEILNSPDRACTLAKQA
FDEAISELDTLGEESYKDSTLIM
QLLRDNLTLWTSDMQDDGGDEMRDATKPEDEH

Nucleotide Sequence for Rice using GeneBank

caagtcgaccactgtaagtacaaccaacagcgggttgcttatctg
cactcggaatccgccaaacctggcagggtatcaccaaaacatcttgctaacttttgacagac
cggtcacttgatctaccagtgcggtggtatcgacaagcgaaccatcgagaagttcgagaagg
ttagtcaatatcccttcgattacgcgcgctcccatcgattcccacgattcgctccctcactcg
aaacacatccattaccccgctcgagtccgaaaattttgcggtgc
gaccgtgatttttctggtggggtatcttaccccgccactcgagtcacggatgcgcttgccctgttcccacaa

aaccttaccaccctgtcgcgcactacatgtcttgcagtcactaaccactg
gacaataggaagccgccgagctcggaaagggt
tccttcaagtacgcctgggttcttgacaagctcaaagccgagcgtga
gcgtggtatcaccattgatatcgctctctggaagttcgaga
ctcctcgctactatgtcaccgtcattggtatgttgtcactgtctcacactatc
atgtattcatcatgctaacatctctc

EMBL Nucleotide Sequence for Rice

```
>embl|AJ252272|AJ252272 Rice tungro bacilliform virus
aagctttattaccatataggcatgatggccataggaattaagggattacataga
ataggtactaaagtaatggtgatgttttatgacgatagtttcggtaaagctaaag
agtataggatcaatagaaatggatatgaatgcaggttgcggagtattttactct
gacttcgctaaatatataaaagacctatcccatcttaagataggaatacaaac
tatgaaaactatgagggaaaaaatctttccatagccataaaaactataggaa
actaatatccaggctagatataaaatgaatgttaaagatatagttgaacaaata
caaggaataactatggttgcacccatggaaatagatccatcacatttagatgg
tggaatttaagcaaattcatgattcaagaaggaacaagtagagttcctaataa
atatatcagaatttgcatggaggagaatcgcttagattttcaaactatacgcaa
atgcacgatccaacagaaataaattctgatgaagatgaagatttaaaaattct
caattaaatgccaaaatggcaacctttcaagaagaaaccctagaacaaaa
ataaaagaagaaaagaaacaacttctagcaaaactagaagctaaaaaga
gcaaaatctttaaatatggcagtcatcgaagatgacttcaacccaaacaacg
gatgattcatattctgaattagaagatctagaattcgaacaattaggattaactg
gaagatctagaccaagaaagtctagaaacagaagaaattactgaatggga
caagtcctaaccagagaaataaaagcctttaaatcagtctctgaacaaatag
tttggagaactattaaaagaacatggaaattatgacatggcccttaagaattta
agatatgatctaaaaaacttagaaggagccaaaaacctagaagaaatagc
acatcaaaaatgatggatgtaaaaccagtaaaaagaccaaaagaagaac
gaagatgatatgagggatgatgggagaagaaaagaattacaacccaatcc
tcaaaagatagaaactttgaaagaataggaggatcatataagaaaaactttt
```

Exercise to Practice

1. Search the protein sequence for maize.
2. Search thc nucleotide sequence for wheat using Gene Bank.
3. Search the nucleotide sequence for maize using EMBL.

FSC
www.fsc.org
MIX
Papier | Fördert gute Waldnutzung
FSC® C083411

Zeitfracht Medien GmbH
Ferdinand-Jühlke-Straße 7
99095 Erfurt, Deutschland
produktsicherheit@kolibri360.de